0.50
74-230

AF731710

ANALYTICAL SPECTROMETRY

ANALYTICAL SPECTROMETRY

L. de Galan

Laboratorium voor Instrumentele Analyse
Technische Hogeschool, Delft

ADAM HILGER LTD
LONDON

This volume is a translation of *Analytische Spectrometrie* which is published and sold in English-speaking and English-reading countries by permission of

N.V. Uitgeversmaatschappij Agon Elsevier
Amsterdam, The Netherlands

ISBN 0 85274 179 0

Published by
ADAM HILGER LTD
Rank Precision Industries Ltd
31 Camden Road, London, NW1 9LP

Printed by
Adlard & Son Ltd, Bartholomew Press, Dorking

PREFACE

The Dutch edition of this book, written in 1969, was based on a course recommended by a committee for a then newly founded type of school more or less comparable to the English technical college. In preparing the English translation errors have been corrected and some sections have been rearranged or rewritten, but basically the contents have remained the same.

This introductory textbook is intended for the general analytical chemist who should be aware of the major spectrometric tools for the qualitative and quantitative analysis of inorganic and organic materials. Consequently, little or no attention is given to spectrometric methods of analysis that have not yet come into general use, such as Raman spectrometry, optical rotatory dispersion and photoelectron spectrometry. For the subjects included, however, I have tried to explain the basic principles and to present the current state of the art emphasizing both the potential and the limitations of each method. The final chapter presents a mutual comparison of the methods on the basis of their analytical characteristics.

There is perhaps no field of analytical chemistry where the lack of a uniform nomenclature and consistent units is more manifest than in spectrometry. I consider this unfortunate and, although I have thought it unwise to negate completely the existence of practical units, I have made a cautious attempt to introduce SI units by making all calculations and writing all equations in this system.

It is a pleasure to acknowledge the cooperation of the publishers, notably Mr N. Goodman and Mr D. Tomlinson in editing this translation.

April 1971 L. DE GALAN

PREFACE

[illegible]

CONTENTS

CHAPTER ONE

ELECTROMAGNETIC RADIATION

1.1. Radiation as a wave phenomenon

Radio and television broadcasting, radar, traffic lights, sunray lamps and X-ray examinations are only a few familiar examples of the many applications of electromagnetic radiation in everyday life. However, for most people the earliest and most frequent contact with electromagnetic radiation is the impression of visible light upon the eye. Although light has been studied for centuries and much is known about its properties, there is no straightforward answer to the simple question: What is light? On the one hand, light is a form of energy, and the production or the absorption of light can only be explained if light is considered to be a stream of energy packets, as will be discussed in the next section. On the other hand, the focusing of light by lenses and its reflection at mirrors can only be explained if light is visualized as a wave phenomenon. This is known as the dual nature of electromagnetic radiation.

In this section the phenomena pertinent to the wave nature of light will be briefly reviewed. In Fig. 1.1 a ray of light moves in the direction of the x-axis. This ray can be represented by an electromagnetic field that varies periodically perpendicular to the direction of propagation. If the electric field varies in the direction of the y-axis, the corresponding magnetic field varies in the direction of the z-axis. This is called a transverse, travelling wave, and it is characterized by the following quantities.

(*a*) The plane determined by the electric wave is called the *plane of polarization*. For the wave drawn in Fig. 1.1 y is the direction of polarization and xy is the plane of polarization. The various waves that together constitute the beam of light generally have different directions of polarization (y, y', y'', . . .); this is called unpolarized light. If all directions of polarization are made to coincide, the result is called a plane-polarized beam of light. Polarization phenomena are of interest in double refraction, rotatory dispersion and Raman spectrometry, but will not be discussed in this book.

(*b*) The maximum value of the electric field strength is called the *amplitude A*. The strength or the *intensity* of the light wave is

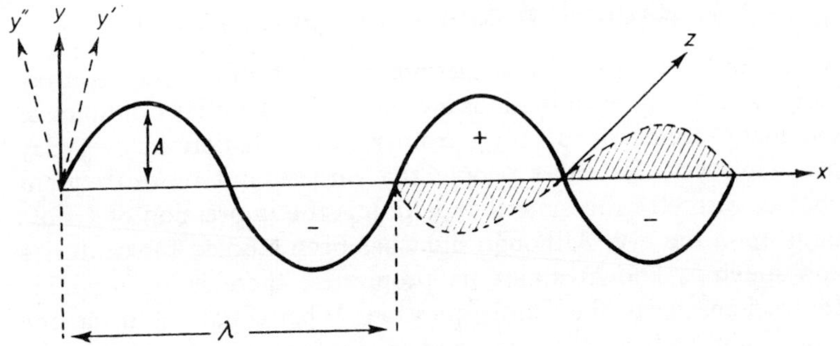

FIG. 1.1. Propagation of waves. The solid line in the *xy*-plane denotes the varying electric field strength. The dashed wave in the *xz*-plane is the varying magnetic field strength. A is the amplitude and λ the wavelength.

proportional to the square of the amplitude. (If E and H denote the periodically varying electric and magnetic field strength, the product $E.H$ in fact has the dimensions of energy passing a cross-section per unit of time, which are just the dimensions of the intensity.)

(*c*) The *frequency* ν denotes the number of times per second that the electric field reaches its maximum positive value.

(*d*) The *wavelength* λ is the distance covered by the wave within one complete oscillation. As will be seen in § 1.4 the wavelength (and the frequency) of electromagnetic waves can vary over many orders of magnitude.

(*e*) Since there are ν oscillations per second and each oscillation covers a distance λ, the product

$$u = \lambda\nu \tag{1.1}$$

denotes the distance travelled by the wave in one second, or the *velocity of propagation.*

The following properties of light are explained by its wave character.

(*a*) In Fig. 1.2a a ray of light of a single wavelength (*monochromatic* radiation) is propagated along a straight line through empty space and strikes another substance at an angle of incidence i with the normal (perpendicular) to the surface. At this point the

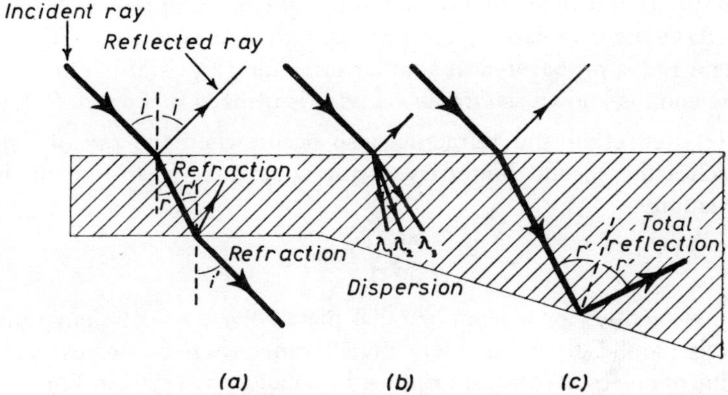

FIG. 1.2. Optical phenomena at the boundary of two media. (a) Refraction and reflection of a monochromatic ray at a plane parallel plate. (b) Dispersion of polychromatic radiation. (c) Total reflection against a medium of low index of refraction.

direction of propagation is changed. One part of the ray is reflected back into space at the same angle i, but to the other side of the normal. The remaining part of the ray enters the substance at an angle of refraction r. The frequency of the waves remains the same, but the direction and the velocity of propagation are altered according to the following law:

$$\frac{\sin i}{\sin r} = \frac{c}{u} = \frac{\lambda_{\text{vacuum}}}{\lambda_{\text{substance}}} = n \tag{1.2}$$

where n is called the ***refractive index*** of the substance (with respect to vacuum). The velocity of propagation in a vacuum, c, is the same for all electromagnetic waves, and this physical constant has the value

$$c = \nu\lambda_{\text{vacuum}} = 2{\cdot}998 \, . \, 10^8 \text{ m/s} \tag{1.3}$$

c is always larger than u and, therefore, $n > 1$. It follows from Equation 1.2 that the wavelength of electromagnetic radiation depends upon the refractive index of the medium. Wavelengths should always be quoted for vacuum, but the difference from the wavelength in air is usually small; for instance for visible light

$$\lambda_{\text{air}} = \lambda_{\text{vacuum}}/n_{\text{air}} = 0{\cdot}9997 \, \lambda_{\text{vacuum}}$$

(*b*) In general, the refractive index is different for each wavelength. If the incident ray contains light of various wavelengths (*polychromatic* radiation), each wavelength is refracted at a different angle and is propagated in another direction (Fig. 1.2b). This phenomenon is known as ***dispersion*** and it is utilized in a prism (§ 3.3).

(*c*) Reflection and refraction also occur when the ray of light reaches the second boundary of the substance, according to the equation

$$\frac{\sin i'}{\sin r'} = n$$

Consequently, for a plane parallel plate $i' = i$, and the emerging ray is parallel to the incident ray. Because $\sin i'$ cannot exceed a value of one ($i' = 90°$), $\sin r'$ must be smaller than $1/n$. In Fig. 1.2c the angle r' is greater than this critical value. In that case the ray is ***totally reflected*** and no radiation emerges from the substance. Total reflection can only occur if the wave travels in a substance of high refractive index (n_1) and reaches a boundary with a medium of lower refractive index (n_2), and the general condition is $\sin r' > n_2/n_1$ (§ 5.4).

(*d*) The wave theory of electromagnetic radiation predicts that when a beam of radiation passes an aperture, the waves will spread in all directions. Indeed, this phenomenon of ***diffraction*** can be observed experimentally, but only if the dimensions of the aperture are of the same order of magnitude as the wavelength of the radiation. For apertures much larger than the wavelength, diffraction only leads to a very slight broadening of the beam.

(*e*) Diffraction of light occurs at each point of the (narrow) aperture. As a result, the waves spreading at each point will mix and cross each other. Now, the diffracted waves of *one and the same* beam give rise to a phenomenon known as *interference.* Let us consider two waves of equal amplitude, that are diffracted at two slightly different angles, so that at some point after the aperture the two waves meet. Now, if the path lengths from the aperture to the point of crossing are exactly equal, both waves will be *in phase* and the electric field strengths reach their maximum values at exactly the same time. The resulting amplitude of the interfering waves is then (Fig. 1.3a)

$$A = A_1 + A_2 = 2A_1 \quad \text{or} \quad I = A^2 = 4A_1^2 = 4I_1$$

However, the diffraction pattern also shows points where one wave is exactly one half of a wavelength ahead of the other wave. The electric field strength of the one wave is then maximally positive

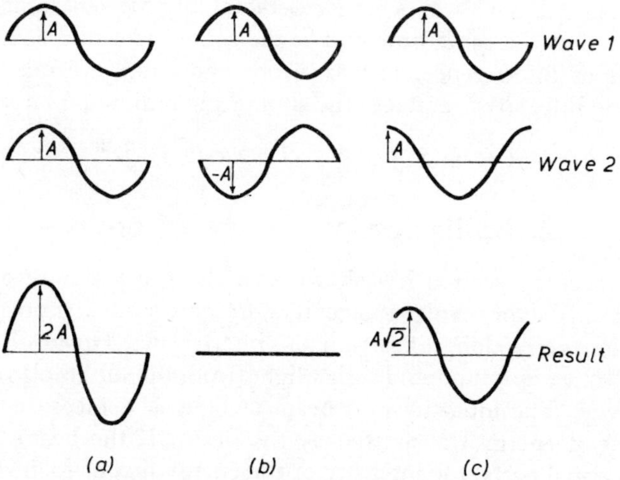

FIG. 1.3. Optical interference. (a) Constructive interference of two waves in phase; path length difference zero or equal to a whole number of wavelengths. (b) Destructive interference of two waves in opposite phase; path length difference equal to a whole number of wavelengths plus one half of a wavelength. (c) Path length difference a whole number of wavelengths plus one quarter wavelength.

when that of the other is maximally negative, and the result is

$$A = A_1 - A_2 = 0 \quad \text{and also} \quad I = 0$$

These two waves are *in opposite phase* and mutually extinguish each other (Fig. 1.3b). At such points the intensity is zero.

If the path lengths travelled by the two interfering waves do not differ by exactly a whole or one half of a wavelength, the result of the interference is more complicated (Fig. 1.3c). In the case of interference the resulting intensity at a particular point cannot be predicted in general, but must be derived from an addition of all contributing waves, each with its proper phase. Diffraction and interference are essential concepts in understanding the operating principle of gratings (§ 3.4).

(*f*) Now, what happens if we mix two arbitrary waves, for example two waves from different sources or from different points of the one light source? It appears that (at least for light waves) such waves do not produce interference patterns. This is expressed by saying that such waves are *incoherent*. Only *coherent* light waves generated at the same time and at the same point of a source are capable of interference. If two incoherent waves are mixed, the resulting intensity is equal to the sum of the individual intensities:

$$I = I_1 + I_2 \quad \text{or} \quad A^2 = A_1^2 + A_2^2$$

1.2. Radiation as a form of energy

The preceding section has shown that the wave picture of light gives a satisfactory explanation of its *optical* properties: propagation, reflection, refraction, diffraction and interference. However, we all know from a hot summer day that light (from the sun) is also a form of energy. The intensity of a beam of light is a measure of the amount of energy transported by the beam. If the beam is split in two equal parts, the intensity or the energy flux of each secondary beam will be half the original value. Now, the question may be asked whether the energy can be reduced to an arbitrary small value by repeating this splitting again and again. If light were only waves this would obviously be possible, because the amplitude of even the weakest wave can always be reduced further and further.

It appears, however, that the answer to the above question is negative. In order to understand this, a beam of monochromatic radiation can best be compared with an electrical current. The electrical current consists of a large number of moving *units of charge*. In the same way, the beam of radiation consists of moving *units of energy*, which are called *photons*. (Note, however, that whereas the units of charge are carried by particles, notably electrons, the photons move on their own.) Photons may be destroyed (absorption) or created (emission), but photons cannot be split. Thus, if we repeatedly split a beam of radiation in two parts, we end up with a single photon with an energy

$$E = h\nu = hc/\lambda_{\text{vacuum}} = hc\sigma_{\text{vacuum}} \qquad (1.4)$$

where h is Planck's constant and has the value

$$h = 6{\cdot}625 \, . \, 10^{-34}\,\text{J s}$$

The energy of the photon is proportional to the frequency and inversely proportional to the (vacuum) wavelength of the radiation. The reciprocal of the wavelength is called the *wavenumber* σ, which is therefore also proportional to the photon energy. Since the frequency of electromagnetic radiation varies over many orders of magnitude (see Fig. 1.4) there must be many photons of widely different magnitude.

Some quite familiar phenomena illustrate the important relation between the frequency of radiation and the energy of individual photons. Radiation of an infra-red lamp warms our skin and heats our food. Radiation of somewhat larger frequency is 'visible', because it produces certain chemical reactions in the retina of the eye. Ultra-violet radiation damages the eye and changes the colour of the skin. High-frequency X-rays have sufficient energy to penetrate the skin and the soft flesh, but not the bones. The extremely energetic γ-rays are lethal. *These different properties are not related to the total energy of the beam, but result entirely from the differences in energy of the basic unit in the beam, the photon.* Even the most intense sunray lamp (ultra-violet radiation) will never give an X-ray photograph!

The photon nature of electromagnetic radiation is most clearly demonstrated by the phenomena related to the emission, the absorption and the detection of radiation. As spectrometric

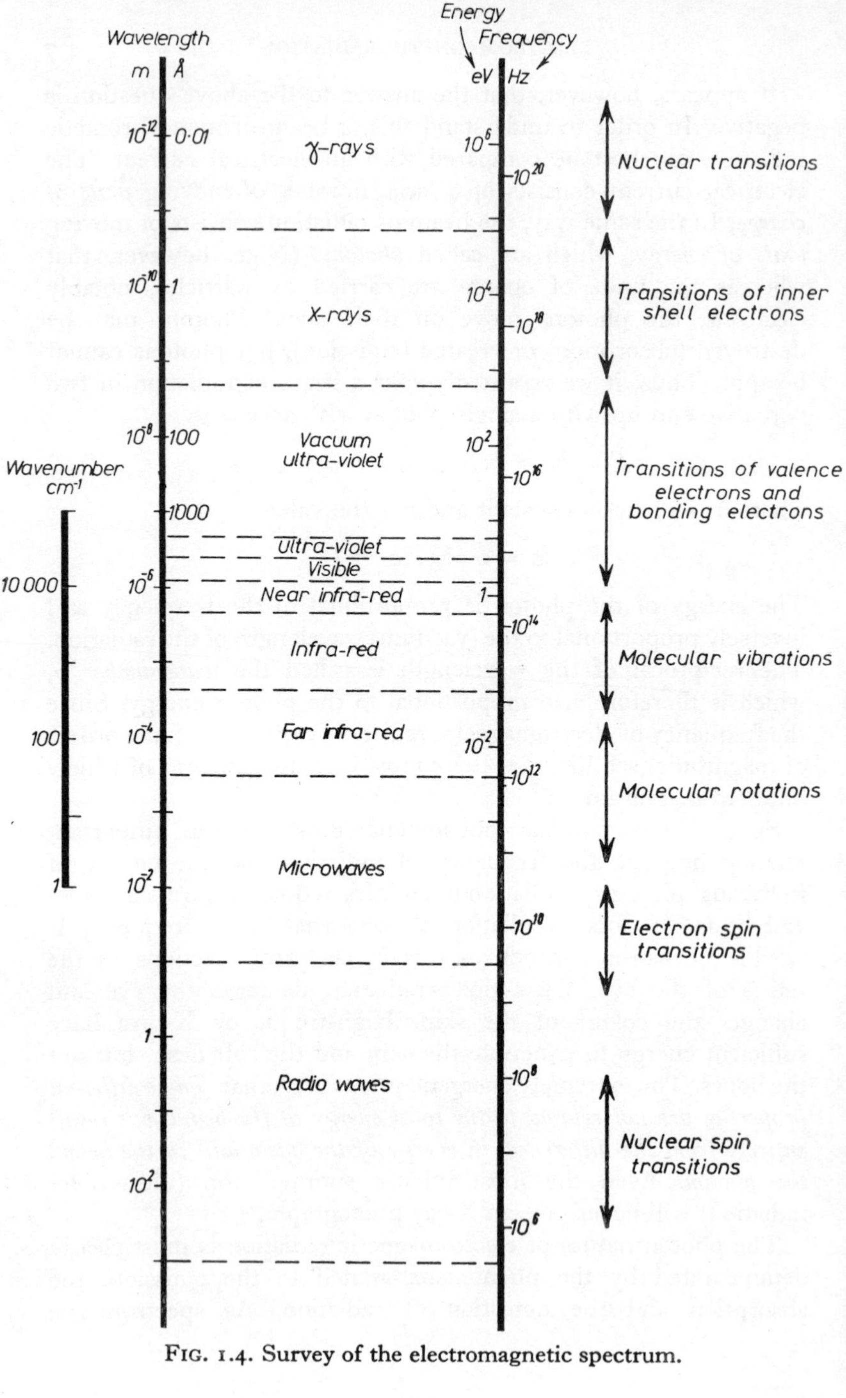

FIG. 1.4. Survey of the electromagnetic spectrum.

methods of analysis are based upon precisely this interaction between electromagnetic radiation and matter, the photon nature of radiation is of fundamental importance for spectrometry.

If radiation is absorbed, energy is removed from the beam of radiation and the absorbing substance gains energy. This produces a change within the substance, that can frequently be related to changes in the molecules, the atoms or the electrons of the substance. Since the beam of radiation consists of photons, and photons cannot be split, the energy of the substance must increase by the energy of one photon, or multiples thereof. The gain and the loss of radiation energy is thus restricted to units of a particular magnitude, and these processes are therefore said to be *quantized*. If the substance originally has an energy E_1, its energy will be increased by the absorption of one photon to

$$E_2 = E_1 + h\nu \tag{1.5}$$

Hence $$h\nu = E_2 - E_1$$

Absorption of radiation is an example of a *spectral transition* or *radiative transition* between two *energy levels* E_1 and E_2 of the substance.

1.3. Some common units

As can be seen in Fig. 1.4, the photon energy, the frequency and the wavelength of electromagnetic radiation vary over an extremely wide range of values. Because simple whole numbers are preferred for daily use and in written texts, a variety of units has been adopted for the various parts of the electromagnetic spectrum. Only the more important ones will be considered here; some other (practical) units will be mentioned in the appropriate chapters.

TABLE 1.1. Recommended prefixes

Prefix	*Name*	*Factor*	*Example*
p	pico	10^{-12}	1 pF = 10^{-12} farad
n	nano	10^{-9}	1 nm = 10^{-9} metre
μ	micro	10^{-6}	1 μA = 10^{-6} ampere
m	milli	10^{-3}	1 mV = 10^{-3} volt
k	kilo	10^{+3}	1 keV = 10^{3} electronvolt
M	mega	10^{+6}	1 MHz = 10^{6} hertz

The recommended unit of length is the metre. Accordingly, the wavelength is preferably expressed in fractions or multiples of the metre by means of the prefixes presented in Table 1.1. In the small-wavelength region (ultra-violet and X-ray) the ångstrom is still widely used:

$$1\ \text{Å} = 10^{-10}\ \text{m} = 0{\cdot}1\ \text{nanometer (nm)}$$

However, in view of the inverse relationship expressed by Equation 1.4 the wavelength is an inconvenient measure for the photon energy, which is the fundamental quantity for analytical spectrometry. Therefore, three other quantities are frequently used to specify electromagnetic radiation, that are either a direct measure for the photon energy or proportional to it.

The recommended unit for energies, the joule, is much too large to be suitable for photon energies, but for wavelengths up to 1 μm the electronvolt is a convenient unit. The electronvolt is defined as the energy of an electron that has been accelerated by a potential difference of one volt. As the charge of an electron is $1{\cdot}6.10^{-19}$ C, 1 eV is equal to $1{\cdot}6.10^{-19}$ V C = $1{\cdot}6.10^{-19}$ J. Using Equation 1.4 we find that, for example, the photon energy of radiation with a (vacuum) wavelength of 1 Å is given by

$$E = 6{\cdot}625.10^{-34} \times 2{\cdot}998.10^{8}/1.10^{-10} = 1{\cdot}986.10^{-15}\ \text{J}$$

$$= 12{\cdot}4\ \text{keV (kilo-electronvolt)}$$

The unit of frequency is the reciprocal second. In English and American spectrometric literature this unit has long been called the cycle or the cycle per second. The European and recommended name is the hertz (Hz), which appears to be gradually becoming accepted and will be used in this book. The frequency has the advantages of being directly proportional to the photon energy and independent of the refractive index of the medium (§ 1.1). However, Fig. 1.4 shows that acceptable numerical values only occur in the long-wavelength region ($\lambda > 1$ cm), when the frequency is given in megahertz (MHz).

In the middle portion of the electromagnetic spectrum the wavenumber, the reciprocal of the vacuum wavelength, is used as it is also proportional to the photon energy (Equation 1.4). Wavenumbers are expressed in a rather unfortunate unit, the reciprocal centimetre. A wavenumber of 1000 cm^{-1} means that there are a thousand waves present in one centimetre and hence the wavelength is 10^{-3} cm or 10 μm. The corresponding photon energy is 2.10^{-20} J = 0·124 eV. This relationship between the wavenumber and the photon energy is often tacitly extended to other energy processes. Indeed, the energy of electrons in atoms and molecules is frequently stated in cm^{-1}, although the dimensions of this unit are clearly not those of energy.

Conversion factors between units with the same dimensions and between units of different dimensions are presented in Appendix II.

1.4. The electromagnetic spectrum

The entire range of electromagnetic waves is called the electromagnetic spectrum, which is shown in Fig. 1.4 and can be roughly divided into the following regions.

(*a*) γ-radiation with wavelengths between 0·001 and 0·01 nm result from changes within the atomic nucleus. This radiation is utilized in neutron activation analysis (Chapter 9).

(*b*) X-rays are emitted when an electron recedes to a vacancy in one of the inner electronic shells of an atom (§ 2.2). The photon energy ranges from 1 to 100 keV, the wavelength from 0·01 to 1 nm. X-rays are applied in qualitative and quantitative inorganic analyses and in determining crystal structures (Chapter 8).

(*c*) Transitions of the valence electrons in atoms and of the bonding electrons in molecules give rise to radiation with a wavelength between 10 and 1000 nm (photon energy 1 to 100 eV). A minor portion of this range is the visible region with wavelengths between 400 and 800 nm. Because oxygen and nitrogen molecules absorb radiation with wavelengths below 200 nm, all air must be removed from instruments used to study the 'vacuum' ultra-violet region. Applications of the ultra-violet and the visible part of the electromagnetic spectrum are discussed in Chapters 4, 6 and 7.

(*d*) Photons with an energy of about 0·1 eV or a wavenumber of 1000 cm^{-1} can produce changes in the oscillatory movement of atoms in molecules. The vibration spectra in the infra-red region (1–25 μm) are more important for analytical purposes than the rotation spectra in the far infra-red (25–1000 μm) that result from changes in the speed of revolution of the molecules around their axes (Chapter 5).

(*e*) In a strong magnetic field the spins of free electrons in a molecule can assume different positions with respect to the magnetic field strength, which results in various energy levels. The radiative transitions between these levels fall in the region of microwaves (1–100 mm), also known from radar techniques. Electron paramagnetic resonance spectrometry is used for the study of free radicals.

(*f*) As the atomic nucleus is much heavier than the electron, the radiative transitions between energy levels of nuclear spins in magnetic fields are shifted towards longer wavelengths (radio waves). Nuclear magnetic resonance spectrometry is the subject of Chapter 10.

1.5. Radiative processes

Let us consider a system of a great number of mutually independent particles, such as the molecules in a diluted solution or the atoms in a gas. The particles are continuously moving (Brownian motion) and, therefore, possess a certain kinetic energy. For all practical purposes the kinetic energy can be considered as non-

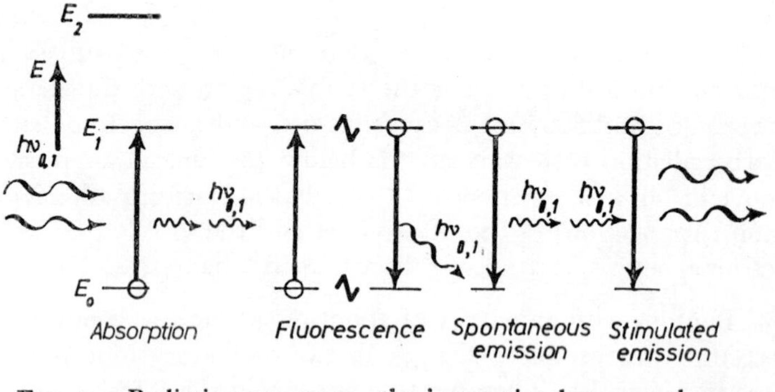

FIG. 1.5. Radiative processes—the interaction between photons and matter.

quantized; that is, the kinetic energy can assume every possible value.

In addition, each particle possesses a potential energy that is related to the position of the atoms in a molecule or to the distribution of the electrons in atoms in molecules. As will be seen in the next chapter, this potential energy cannot assume any arbitrary value, but is restricted to certain discrete values. If the potential energy is plotted on a vertical axis, these discrete values correspond to fixed points on the energy scale. If horizontal lines are drawn through these points, as is done in Fig. 1.5, the resulting potential *energy levels* together form an *energy level diagram*. Each particle of the system can only possess a potential energy indicated by one of the levels in the diagram. Intermediate values of potential energy are impossible.

Now, if the substance is irradiated with electromagnetic radiation, the particles can react with the individual photons in the beam. If, as a result of this interaction, the potential energy of a particle is increased by the energy of a photon, the radiative process is known as *absorption*. Because the particle can only jump from one energy level to another energy level and since, as we saw in § 1.2, a photon cannot be split, absorption of radiation is obviously only feasible if the photon energy agrees exactly with the difference in energy between two levels drawn in Fig. 1.5. By absorption of a photon with frequency $\nu_{0,1}$ and energy $h\nu_{0,1} = E_1 - E_0$ the particle is excited from a lower energy level E_0 to a higher energy level E_1.

What does the particle do with this extra potential energy? Suppose that it returns to the lower level E_0 immediately and emits the photon again. In that case the original attenuation of the incident beam of radiation will be compensated immediately and it is simply not possible to measure any absorption at all. Fortunately, this does not happen. In general, the particle loses its extra energy through collisions with other particles, such as solvent molecules. This is called a *radiationless* transition between the energy levels E_1 and E_0. The potential energy gained through absorption of radiation is then converted into kinetic energy and, in fact, the photon energy of the incident beam is converted into heat.

However, occasionally this conversion of energy does not occur.

The particle keeps its additional potential energy for some time (generally a very short time, a few millionths of a second) and re-emits a photon with an energy $h\nu_{0,1}$. This is called *fluorescence* (Fig. 1.5). However, even this short time lag between absorption and re-emission is long enough to let the particle 'lose its sense of direction'. As a result the fluorescence radiation can be emitted in any arbitrary direction and the fluorescence radiation of the substance as a whole can be detected separately. Examples will be considered in § 4.8, 6.5 and 8.5.

Let us now return to Fig. 1.5. At low temperatures (in extreme cases at absolute zero) each particle will possess the minimum possible potential energy. All particles are then in the *ground state* E_0. If now the substance is heated, the kinetic energy of the particles increases and the reverse process can take place: through collisions some particles convert kinetic energy into potential energy, leave the ground state E_0 and are excited to one of the higher energy levels $E_1, E_2, \ldots$. Just as in the case of fluorescence, it is now possible that the particle returns from the excited level E_1 to the ground state E_0 by emission of a photon of energy $h\nu_{0,1}$. This phenomenon of ***spontaneous emission*** is responsible for the emission of light by light bulbs and other sources of radiation. Electrical or chemical energy is converted through kinetic and potential energy of the particles into electromagnetic, radiative energy. Applications of spontaneous emission in analytical spectrometry are considered in Chapters 6 and 7.

Finally, there is a fourth radiative process possible. The particle may also return from the excited state E_1 to the ground state E_0 under the influence of a beam of photons of the correct energy $h\nu_0$. Now, instead of being attenuated by absorption, the intensity of the beam is increased by ***stimulated emission***. Thus, photons with an energy $h\nu_{0,1}$ are not only absorbed by particles in the ground state E_0, but will also be augmented by particles in the higher energy level E_1. ***The net result of these two competing processes depends only upon the relative numbers of particles in the states* 0 *and* 1.**

It has been shown by Boltzmann that for a substance with a homogeneous temperature of T°K the following equation holds:

$$\frac{n_s}{n_0} = \frac{g_s}{g_0} \exp \frac{-(E_s - E_0)}{kT} \qquad (1.6)$$

Here n_s and n_0 are the numbers of particles in the higher state s with an *excitation energy* E_s and in the ground state with an energy E_0, respectively; g_s and g_0 are so-called weight factors, which are usually of the same order of magnitude, and k is a physical constant (Boltzmann's constant) with a value

$$k = 1{\cdot}38.10^{-23}\ \mathrm{J/°K}$$

It follows from Boltzmann's equation that there are always more particles in the ground state than in a higher energy level ($E_s > E_0$) and, therefore, when a substance is irradiated with

TABLE 1.2. The population of excited energy levels according to Boltzmann's equation

Field of analysis	*Photon wavelength*	*Photon energy* (*eV*)	*Wave-number* (*cm*$^{-1}$)	$\exp[-(E_s-E_0)/kT]$ 300°K	2500°K	5500°K
NMR	5 m	10^{-7}	0·002	0·9999	$1-10^{-6}$	
EPR	0·1 m	10^{-5}	0·1	0·9995	$1-10^{-5}$	
Infra-red	1000 μm	0·001	10	0·95	0·995	0·998
	100 μm	0·01	100	0·62	0·94	0·97
	10 μm	0·12	1000	8.10^{-3}	0·56	0·78
	5 μm	0·25	2000	7.10^{-5}	0·32	0·60
	2 μm	0·62	5000	10^{-10}	0·06	0·28
Visible	1 μm	1·24	10000	—	3.10^{-3}	0·08
	500 nm	2·5	20000	—	1.10^{-5}	7.10^{-3}
	400 nm	3·1	25000	—	6.10^{-7}	2.10^{-3}
Ultra-violet	300 nm	4·1	33333	—	5.10^{-9}	2.10^{-4}
	200 nm	6·2	50000	—	3.10^{-13}	3.10^{-6}

electromagnetic radiation the absorption will always predominate over the stimulated emission.* However, it is also clear that the relative number of particles in the higher level (n_s) becomes larger if the temperature T increases or if the excitation energy E_s decreases. This can be seen from the figures in Table 1.2, where the

* Boltzmann's equation is quite valid for all systems encountered in this book. However, the *laser* forms an example of a system that does not obey Boltzmann's equation. In this device the relative population of two higher energy levels ($E_2 > E_1$) is reversed under the influence of an extremely intense beam of electromagnetic radiation of photon energy $h\nu_{0.2} = E_2 - E_0$ (saturation, see § 10.2). Consequently, $n_2 > n_1$ and between the levels E_2 and E_1 the stimulated emission predominates over the absorption, which leads to the remarkable properties of the laser.

exponential factor has been calculated for a few temperatures and for several excitation energies corresponding to photon energies of radiation of varying wavelength.

For radiation applied in nuclear magnetic resonance and electron paramagnetic resonance (wavelength from 0·1 to 10 m) the photon energy and, hence, the energy difference $E_s - E_0$ is so small that at room temperature both levels are nearly equally populated. For every 10 000 particles in the ground state there are 9999 particles in the higher state. For an incident beam of this wavelength there are 9999 stimulated emission transitions for every 10 000 absorption transitions, and the net attenuation of the beam is extremely small. This will be further discussed in Chapter 10.

As the wavelength decreases, the relative population of the higher energy level decreases, and in the infra-red region stimulated emission can already be neglected at room temperature. The very low population of higher energy levels explains why the objects around us do not emit visible radiation at room temperature (wavelength from 400 to 800 nm). Strong heating in flames (T = 2500°K) or in electrical discharges (5500°K) is required to raise the population to such a degree that spontaneous emission of radiation can be observed (Chapters 6 and 7).

Radiation of very small wavelength cannot be excited by thermal heating, but can only be produced by bombarding the substance with electrons, neutrons or X-rays (Chapters 8 and 9).

1.6. Analytical spectrometry

The applications of electromagnetic radiation in solving analytical problems can now be readily understood. The set of the energy levels forms a unique property of a substance, and their mutual separation determines the energy and the wavelengths of the electromagnetic waves that can be absorbed or emitted by that substance. In turn, if we measure these wavelengths, we obtain information about the nature, the structure and the composition of our sample. This is the basis of qualitative analysis in its broadest sense.

It has also been remarked that the intensity of spontaneous emission or the extent of absorption depends upon the number of particles in the relevant energy levels. Therefore the emission or the absorption intensity is a measure of the amount of the component responsible for a particular transition. This is the basis of quantitative analysis.

Qualitative and quantitative information is obtained through the *spectrum*. The absorption or emission spectrum of a substance is the collection of electromagnetic waves that can be absorbed or emitted by the substance, each with its proper intensity. Such a spectrum may be represented as a drawing, such as Figs. 2.3 and 2.4, as a photograph, such as Plate 1, or simply as a table listing numerical values of wavelengths and corresponding intensities. However, it should be realized that even a simple compound, such as NO, can under suitable conditions absorb or emit radiowaves, microwaves, infra-red and ultra-violet radiation, X-rays and γ-rays. Thus, the complete spectrum of a substance covers all parts of the electromagnetic spectrum. Generally, we are only interested in that portion of the spectrum that will give us the information we seek. Therefore, spectra are subdivided. One broad subdivision stems from the nature of the radiative process considered, and we speak of absorption, emission or fluorescence spectra. A second classification is derived from the wavelength region studied, such as ultra-violet absorption spectrum, X-ray fluorescence spectrum, etc. Finally, the spectrum may be indicated by the particles or by the process involved in radiative interactions: examples are the nuclear magnetic resonance spectrum, molecular vibration spectrum, atomic emission spectrum, etc. These terms will be used in the following chapters.

Between the radiation process and the analytical results stands the instrument used to detect the radiation and to record the spectrum. Such instruments will be called *spectrometers*. The quality of an analytical result depends upon the quality of the spectrometer, the skill and care of the operator, the composition of the sample and the properties of the component that must be determined. Each of these factors is a potential source of errors varying from slight irregularities to large deviations that will render the result less reliable.

This introductory chapter may be concluded with a comprehensive list of terms that are commonly used to characterize an analytical determination. These terms will also be used extensively in the following chapters.

(*a*) The *precision* of a determination is determined by the *random* error in the measured signal. Standard formulae are used to calculate the standard deviation, the average, and the standard deviation of the average from a series of results. Unless otherwise stated the precision quoted in this book will always refer to the standard deviation or the coefficient of variation (which is equal to the standard deviation divided by the concentration).

(*b*) The *accuracy* of a determination denotes how closely the experimental result agrees with the 'true' answer, and indicates the presence or absence of *systematic* errors. Whereas it is always possible to find an estimate of the precision, it is not always possible to assess the accuracy of a determination. The usual procedure is to analyse a reference sample of well known composition and check the results. It should be emphasized that a very precise determination can be subject to an appreciable unknown systematic error. In fact, this occurs only too often in spectrometric analyses.

(*c*) One frequent source of systematic errors is the presence of other components in the sample. Such components are then said to *interfere* with the determination.

(*d*) The result of a quantitative analysis is often read from an *analytical curve*, which is a graphical representation of the measured signal x as a function of the concentration c (see Fig. 4.6).

(*e*) The *sensitivity* S of a determination is defined as the change in the measured signal produced by a small variation of the concentration of the component that is being determined:

$$S = \frac{dx}{dc} \tag{1.7}$$

For a linear analytical curve the sensitivity is equal to the slope of the analytical curve, and independent of the concentration. For a non-linear analytical curve the sensitivity varies with the concentration.

(*f*) The *background* is the signal due to all other factors besides the component that is determined. Cosmic radiation forms a background in measuring radioactivity. The flame spectrum produces background radiation in flame emission spectrometry. Quite often other constituents of the sample contribute to the background. Therefore, a *blank* determination is usually carried out with a sample that does not contain the component of interest. The corresponding signal is the blank signal x_b.

(*g*) *Noise* is a term derived from electronics to indicate the *rapid*, irregular fluctuations of the measured signal (compare Fig. 8.8b). Noise is an important factor in determining the precision.

(*h*) *Drift* is the slow, fairly regular change of the signal over an extended period of time.

(*i*) The *limit of detection* denotes the smallest amount or the lowest concentration of a component that can be detected.

It is clear that small concentrations approaching the limit of detection will give rise to small signals only, and some criterion must be adopted to distinguish these signals from the random fluctuations of varying magnitude (the noise) of the background. For sound statistical reasons the commonly accepted criterion is that a signal is considered to be genuine or *significant* if it is equal to or larger than

$$x_{min} = x_b + 3s_b \tag{1.8}$$

Here x_b is the average blank signal and s_b is the standard deviation of repeatedly measured blank signals. The limit of detection is now defined as the concentration corresponding to the signal x_{min}. Since the blank signal corresponds to zero concentration, combination of Equations 1.7 and 1.8 yields for the limit of detection,

$$c_{min} = \frac{x_{min} - x_b}{S} = \frac{3s_b}{S} \tag{1.9}$$

Bibliography

1. JENKINS, F. A., and WHITE, H. E., *Fundamentals of Optics*, McGraw-Hill, New York, 3rd ed., 1957.
2. TENQUIST, D. W., WHITTLE, R. M., and YARWOOD, J., *University Optics*, Iliffe, London, 1969.
3. BAUER, E. L., *A Statistical Manual for Chemists*, Academic Press, New York, 1960.

4. ECKSCHLAGER, K., *Errors, Measurement and Results in Chemical Analysis*, Van Nostrand, London, 1969.
5. DOERFFEL, K., *Beurteilung von Analysenverfahren und -Ergebuissen*, Springer, Berlin, 1962.

Problems

1. A beam of yellow sodium light with a vacuum wavelength of 589 nm strikes the following substances at an angle of incidence of 45°.

Substance	*Refractive index for λ_{vacuum} = 589 nm*
Air	1·0003
NaCl	1·544
KCl	1·490
Glass	1·627
Water	1·333

Calculate the velocity of propagation and the wavelength in these substances, and the angle of refraction.

2. Two light waves of amplitude A, emitted at the same time from the same point of a light source, have travelled along different paths and when they meet they are $\frac{1}{8}$ wavelength out of phase. What is the resulting amplitude after interference?

3. Calculate the frequency in Hz, the wave number in cm^{-1}, and the photon energy in eV for radiation of the following wavelengths: 1 km, 2·5 m, 0·4 cm, 20 μm, 500 nm, 150 Å and 0·1 Å.

4. The excitation energies and the weight factors of two levels n and m are E_n, g_n and E_m, g_m, respectively. Use Equation 1.6 to derive an expression for the ratio of the numbers of particles populating these levels at a temperature T. Calculate the ratio for the following data: $E_n = 2{\cdot}3$ eV, $E_m = 3{\cdot}5$ eV, $g_n = 2$, $g_m = 4$, $T = 6000$°K.

5. The signal of an analytical determination is read from an electrical recorder. The blank signal is 1·2 mV with a standard deviation of 0·3 mV, whereas a signal of 8 mV is measured for a concentration of 0·05 millimol/l. If it is assumed that the analytical curve is linear, calculate the sensitivity and the limit of detection of this determination.

6. Consider a system of a thousand particles. For each particle two energy levels are available with an 'energy' separation of 230 cm^{-1}; the weight factors of both levels are equal. Calculate the energy in eV, and the frequency and the wavelength of the photons that can be absorbed by this system. How many particles are available for absorption at room temperature, and how many for induced emission?

CHAPTER TWO

ATOMIC AND MOLECULAR SPECTRA

2.1. The spectrum of the hydrogen atom

The ultra-violet, the visible and the infra-red region of the electromagnetic spectrum are often collected under the name *optical region*. This central portion of the electromagnetic spectrum is the most fruitful one for analytical spectrometry and forms the subject of the next five chapters of this book. As an introduction, the origin and the characteristic features of optical spectra will be discussed in this chapter.

The spectra emitted by free atoms in a flame or in an electrical discharge are conveniently recorded with a spectrograph (§ 7.3). As will be seen in the next chapter, each emitted frequency gives rise to a thin black line on the photographic plates. An example of a typical atomic spectrum is shown in Plate 1a. Although for some elements the number of *spectral lines* may run into the thousands, it appears that a free atom does not emit radiation of all frequencies, but can only emit radiation of fixed wavelengths. Accordingly, such a spectrum is called a discrete line spectrum.

At the turn of the century the atomic spectra of many elements had been recorded and the characteristic frequencies or wavelengths had been measured with extreme precision. However, all attempts to explain the origin of these spectra with the theoretical knowledge then available failed miserably. In fact, the challenge posed by the atomic spectra led to a revolutionary change in the concepts of atoms and molecules, and out of this grew an entirely new branch of theoretical physics now known as quantum mechanics. The first step in this direction was taken in 1901 by Planck, who proposed the concept of quantization of radiation energy and introduced the constant h, which was subsequently named after him. The quantized photon energy $h\nu$ formed the basis of Einstein's explanation of the photoelectric effect a few years later (§ 4.3). The current model of the atom was born in 1911, when Rutherford showed that the mass of an atom is concentrated in a tiny, positively charged nucleus, surrounded by a cloud of negative electrons.

The theory of atomic and molecular spectra was founded in 1913, when Bohr advanced the hypothesis that the average distance of an electron from the atomic nucleus cannot take any arbitrary value but is limited to certain fixed values. This is usually expressed by saying that the electron revolves around the nucleus in certain *orbits*, similar to the motion of the planets around the sun. Each orbit corresponds to and can be denoted by a discrete value of the potential energy of the electron (in this chapter the kinetic energy of the particles can be ignored and the term energy will be used throughout to denote the potential energy). Radiation can only be emitted if the electron jumps from a faraway orbit of high energy to an orbit of lower energy that is closer to the nucleus. The energy difference is then expelled by the emission of a photon.

With a few other assumptions Bohr succeeded in calculating the orbits of the electron in a hydrogen atom, and the corresponding orbital energies could be expressed as

$$E_n = -\frac{e^4 m}{8\epsilon_0^2 h^2} \cdot \frac{1}{n^2} \qquad (2.1)$$

Here e and m are the charge and the mass of the electron, h is Planck's constant and ϵ_0 is a constant resulting from the fact that the equation is written in SI-units (Appendix I). Apart from these constant factors, the energy of the electron in the hydrogen atom is solely determined by the integer n, which is called the *principal quantum number* and which can assume only positive, whole values ($n = 1, 2, 3, \ldots$).

The minimum energy results when $n = 1$; this is the ground state of the hydrogen atom. Substitution of the values given in Appendix I for the constants, yields

$$E_{n=1} = -21{\cdot}8.10^{-19}\ \text{J} \quad \text{or} \quad E_{n=1} = -109700\ \text{cm}^{-1}$$

With increasing value of *n* the average distance of the electron from the nucleus and hence its energy increases. At infinitely large *n* the electron is completely separated from the nucleus and the atom is ionized; the electron's energy has then become zero.

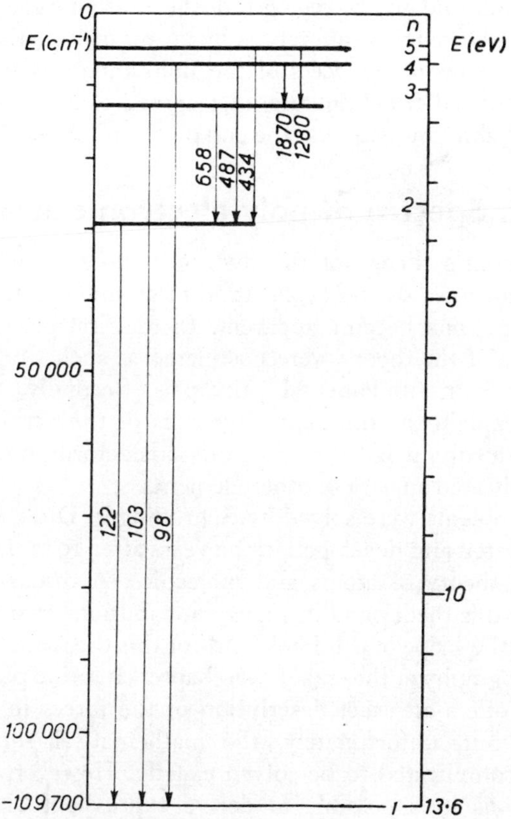

FIG. 2.1. Energy level diagram of the hydrogen atom. The energy levels for low quantum numbers are shown ($n = 1$ to 5); for some spectral transitions the wavelength is indicated in nm.

Consequently, the *ionization energy* that must be supplied to ionize the hydrogen atom from the ground state is equal to $+109700\ cm^{-1}$ or $+13{\cdot}6$ eV.

Fig. 2.1 presents the *energy level diagram* of the hydrogen atom. The horizontal solid lines indicate the discrete energy levels calculated from Equation 2.1 for principal quantum numbers up to five. The vertical arrows represent spectral transitions due to the electron jumping from a high energy level to a lower energy level. The frequency and the wavelength of the emitted radiation can be derived from the energy differences by means of Equations 1.3 and 1.5. It was an historic success of the Bohr theory that the wavelengths calculated from Equation 2.1 agreed exactly with the experimental values measured at the end of the nineteenth century.

2.2. Spectra of poly-electronic atoms

Although Bohr's theory of the hydrogen atom formed a major breakthrough towards the explanation of atomic spectra, its three shortcomings soon became apparent. In the first place, the basic assumptions of the theory were postulated as such, and could not be derived from fundamental principles. Secondly, the theory could not explain the different intensities of the various spectral lines; and thirdly, it failed to account satisfactorily for the much more complicated spectra of other elements.

These problems were solved by Schrödinger, Dirac and others, who formulated and developed in the years after 1925 the quantum mechanical theory of atoms and molecules. A discussion of the concepts of this theory and its many contributions to spectrometry far exceeds the scope of this book, and for this the reader is referred to the bibliography at the end of this chapter. In principle, quantum mechanics offers an exact description of the forces in atoms and molecules, but, unfortunately, the mathematical equations are much too complicated to be solved exactly. However, with some simplifications it is possible to derive expressions and to draw conclusions that are sufficiently accurate for the purposes of analytical spectrometry. A few results will be given in this chapter, but it must be kept in mind that they are only an approximation to the real situation. Also, the following discussion of atomic and molecular spectra will be presented in terms of Bohr's model of the atom. This model provides a convenient qualitative description of the phenomena, but it should always be realized that it is only a picture that cannot correspond to the actual behaviour

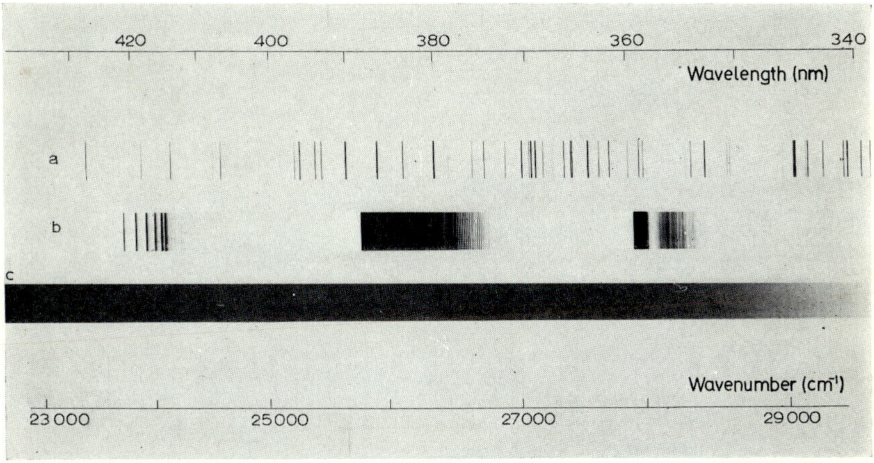

PLATE 1. Different types of emission spectra. (a) Emission spectrum of molybdenum in an electrical spark discharge. (b) Emission spectrum of CN in an electrical arc discharge. (c) Emission spectrum of a light bulb.

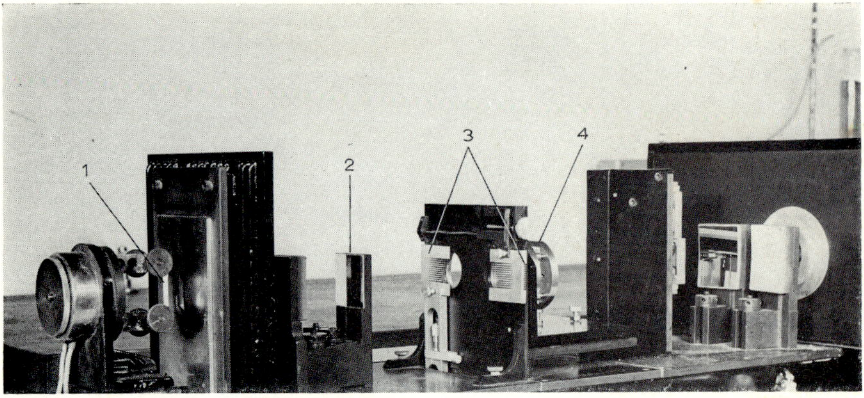

PLATE 2. Detailed view of the entrance optics of an infra-red spectrometer.

1. Source of radiation.
2. Beam splitter.
3. Combs for attenuating the radiation.
4. Absorption cell.

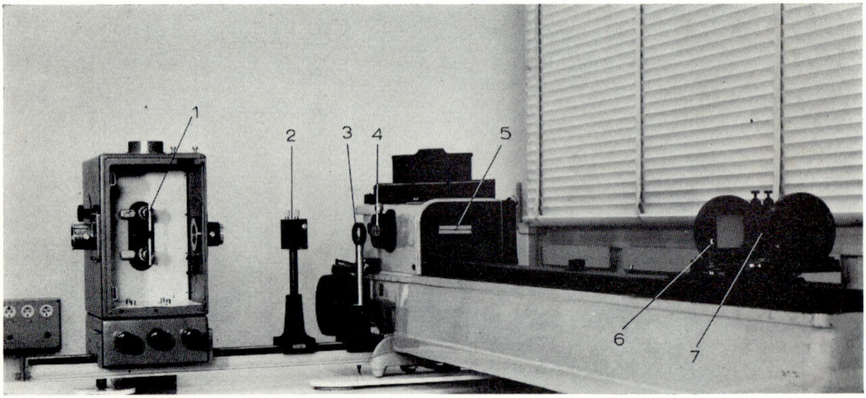

PLATE 3. View of an arc spectrograph.

1. Electrodes.
2. Diaphragm.
3. Lens.
4. Entrance slit.
5. Spectrum.
6. Collimator.
7. Prism with reflecting back surface.

of electrons, atoms and molecules, and should never be taken too literally.

In atoms with several electrons the average position of each electron (the 'orbit') and the corresponding energy is described by four quantum numbers, as follows.

(*a*) In the first instance the energy is determined by the ***principal quantum number*** n, which can assume positive whole values ($n = 1, 2, 3, \ldots$). Orbits with the same value for n belong to the same *shell*, with the following notation:

the K-shell for $n = 1$
the L-shell for $n = 2$
the M-shell for $n = 3$
the N-shell for $n = 4$

and so on (see Chapter 8).

(*b*) In each shell the orbits are distinguished by their *orbital quantum number* l, which can take the values $l = 0, 1, 2, \ldots n-1$. For small values of l the electron approaches the nucleus more closely and the orbital energy is reduced. Again, the various orbits are characterized by letters:

an *s*-orbit or an *s*-electron has $l = 0$
a *p*-orbit or a *p*-electron has $l = 1$
a *d*-orbit or a *d*-electron has $l = 2$
an *f*-orbit or an *f*-electron has $l = 3$

Since l cannot exceed a value of $n-1$, *f*-orbits can only occur if $n \geqslant 4$, corresponding to the N-shell or higher shells.

(*c*) Each electron has a *spin quantum number* s, which is either $+\frac{1}{2}$ or $-\frac{1}{2}$. In the classical picture of the atom the electron not only revolves around the nucleus, but also spins around its own axis. This spinning motion can be either clockwise or counter-clockwise and as a result each orbit with certain values for n and l is split into two orbits with slightly different energies and distinguished by the sign of the spin quantum number s (unless $l = 0$).

(*d*) The *magnetic quantum number* m can assume the values $+l$, $l-1, l-2, \ldots 2-l, 1-l, -l$. Orbits that differ only in the value of m, will only show different energies when the atom is brought into a strong magnetic field. Therefore, this quantum number is of little interest in analytical spectrometry.

However, the magnetic quantum number is important if the electrons are distributed over various orbits in the atom. Each electron will seek to occupy the orbit with the lowest energy, but according to the *exclusion principle* formulated by Pauli, two electrons in *one and the same* atom cannot have an identical set of values for all four quantum numbers. Consequently, the eight electrons of the oxygen atom cannot all occupy the energetically most favourable *K*-shell, because this shell can accommodate only two electrons ($n = 1$, $l = 0$, $m = 0$ and $s = +\frac{1}{2}$ or $s = -\frac{1}{2}$, respectively). The remaining six electrons must take places in the *L*-shell, where there is room for eight electrons: $n = 2$, $l = 0$, $m = 0$, and $l = 1$, $m = -1, 0, +1$, each with $s = +\frac{1}{2}$ or $s = -\frac{1}{2}$. The sodium atom has eleven electrons, and so two can be placed in the *K*-shell, eight can be placed in the *L*-shell and the remaining electron must occupy an orbit in the *M*-shell.

In this way each electron moves in its own orbit characterized by a unique set of values for the four quantum numbers n, l, m and s. Each orbit represents a certain energy. However, the shape of the orbit and the corresponding orbital energy also depend upon the distribution of the other electrons in the atom. It is therefore extremely difficult to calculate the energy of the orbits. Fortunately, completely occupied inner shells (*closed* shells) exert little or no influence upon the energy of the outer orbits. The *valence electrons* in the outer orbits are of special interest to the chemist, because these hold the atoms in a molecule together (*bonding* electrons). Also, the valence electrons determine the absorption and emission spectra in the ultra-violet and visible region of the electromagnetic spectrum. The spectrum of atomic sodium can be completely attributed to the single electron in the *M*-shell.

If the valence electron of sodium is in the lowest orbit of the *M*-shell the sodium atom cannot emit radiation, because all lower orbits are already filled. However, if the atom is heated in a flame or in an electrical discharge, the valence electron can be transferred to other orbits in the *M*-shell or to orbits in the *N*-shell or even further removed shells. From there the electron can recede to its original orbit in the M-shell and this can result in the emission of radiation.

A small portion of the energy level diagram of the sodium atom is shown in Fig. 2.2. The ionization energy required to remove the valence electron from the ground state (in the M-shell) is equal to

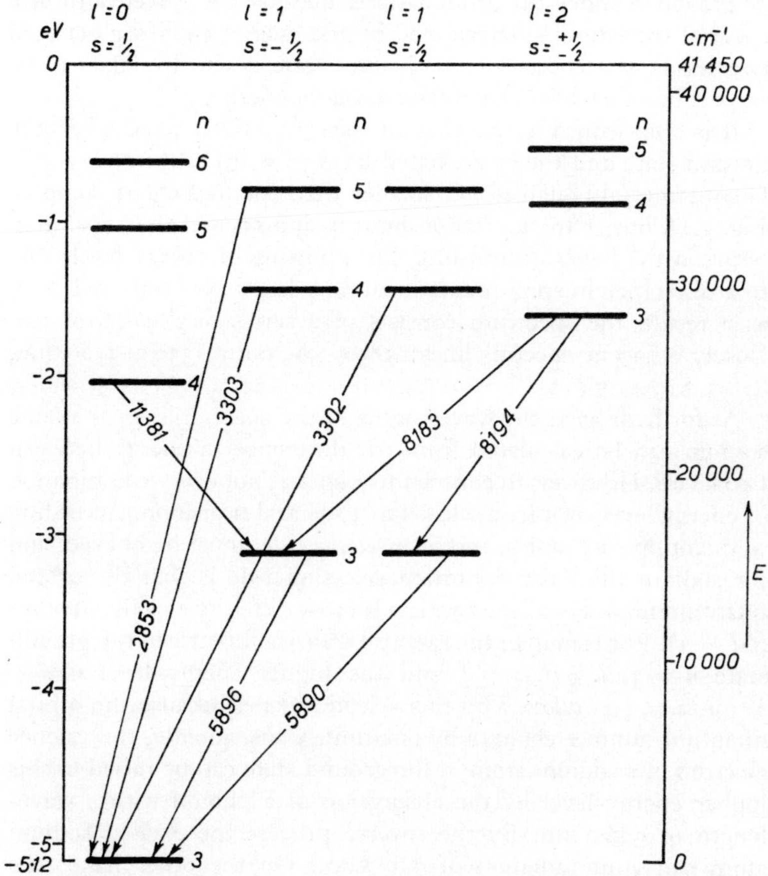

FIG. 2.2. Energy level diagram of the sodium atom. The energy levels are denoted by the values for the principal quantum number n, the orbital quantum number l and the spin quantum number s. Levels with $l = 0$ are not split; for $l = 1$ two separate levels are drawn ($s = +\frac{1}{2}$, $s = -\frac{1}{2}$); for $l > 1$ the splitting is too small to be shown in the figure. Wavelengths of a few spectral transitions are given in Å (1 Å = 0·1 nm).

5·12 eV or 41 450 cm^{-1}. The ground state is characterized by the quantum numbers $n = 3$, $l = 0$, $s = \frac{1}{2}$ (the magnetic quantum number has no influence upon the orbital energy and need not be considered). Higher energy levels are characterized by other numbers. For example, the level with $n = 4$, $l = 1$, $s = +\frac{1}{2}$ lies 30 272 cm^{-1} above the ground state. Because the wavelength of a spectral transition is determined by the energy *difference* between two levels, the energy of the ground state is usually taken to be zero. The other levels then have positive energy.

It is seen from Fig. 2.2 that the energy difference between the ground state and the first excited level ($n = 3$, $l = 1$, $s = -\frac{1}{2}$) is fairly large (although much smaller than for hydrogen; compare Fig. 2.1), but as the ionization limit is approached the separation between the levels diminishes. The splitting of energy levels due to a difference in spin quantum number is always very small and, as a result, the spectrum consists of a series of *doublets* of two closely adjacent spectral lines: 589·0–589·6 nm, 330·2–330·3 nm, 818·3–819·4 nm, etc.

As for hydrogen, the wavelengths of the spectral lines of atomic sodium can be calculated from the difference in energy between two levels. However, in contrast to Fig. 2.1, not every combination of energy levels in Fig. 2.2 leads to a spectral transition. According to quantum mechanics, certain *selection rules* must be obeyed, and for sodium the most important selection rule is that the orbital quantum numbers of the two levels must differ by exactly one unit ($\Delta l = 1$). For example, the spectral transition between the ground state $n = 3$, $l = 0$, $s = \frac{1}{2}$ and the higher energy level $n = 4$, $l = 1$, $s = \frac{1}{2}$ is *allowed* by this selection rule, because the orbital quantum number changes by one unit. Consequently, the valence electron of a sodium atom in the ground state can be raised to this higher energy level by the absorption of a photon with a wavelength of 330·2 nm. By the reverse process the excited sodium atom can emit radiation of 330·2 nm. On the other hand, the *spectral* transition from $n = 4$, $l = 1$, $s = \frac{1}{2}$ to $n = 3$, $l = 1$, $s = \frac{1}{2}$ is *forbidden*, because the orbital quantum number does not change. Therefore if the valence electron jumps between these levels, the energy must be supplied or dissipated in other ways, notably through collisions with other particles. This is known as a *radiationless* transition.

Although the selection rules predict the spectral transitions that may be observed in the emission or in the absorption spectrum, they tell us nothing about the intensity of the various lines. Usually, transitions to and from the ground state give rise to the strongest spectral lines in the spectrum; these are called *resonance lines*. In the *emission* spectrum the resonance transition between the two lowest energy levels is by far the strongest one. The characteristic yellow colour of a sodium street lamp results from the strong resonance lines at 589·0 and 589·6 nm (compare Fig. 4.4).

The energy level diagram of sodium and other alkali metals is comparatively simple as a result of the single valence electron in the outer shell. If there are more electrons in the outer shell, the energy of the atom is no longer equal to the energy of a single electron (plus a constant contribution of the closed shells), but is a function of all electrons in the outer shell. This occurs for alkaline earth elements and especially for the transition elements, such as iron, tungsten, nickel, the rare earths, etc. The energy level diagram becomes quite complicated (for the rare earths the level diagrams are not completely known yet!) and the number of possible combinations of energy levels and hence the number of spectral lines is extremely large. Energy level diagrams of most elements can be found in Candler's book (reference 1).

2.3. Electronic spectra of molecules

Plate 1 shows that the spectrum of even a simple diatomic molecule differs markedly from the spectrum of a free atom. Instead of a number of sharp lines distributed at random over the entire spectrum, we observe a few dark patches which are called *bands*. On closer inspection each band appears to consist of a large number of closely packed lines. The origin of molecular band spectra will be considered in this section.

When two atoms combine to form a molecule, the electrons in the closed inner shells remain with their own nucleus. The valence electrons in the outer shell, however, are strongly influenced by the presence of the other nucleus. Their orbits are reshaped and some electrons describe orbits around and between both nuclei. These are the *bonding electrons* and the orbits are appropriately called *molecular orbits*. Under normal conditions, the bonding electrons occupy the molecular orbits of lowest energy (with due consideration of the exclusion principle), but absorption of radiation energy or heating can raise a bonding electron to an orbit (level) of higher energy. However, the number of excited energy levels in molecules is rather limited and hence the number of possible electronic spectral transitions in a molecule is very much smaller than in a free atom. The energy difference between two levels is some ten thousand cm^{-1}, and this determines the main position of the molecular band in the spectrum (§ 4.4).

In the description of atomic spectra the influence of the atomic nucleus has been ignored, which is perfectly allowable for analytical purposes.* In molecular spectra the influence of the atomic nuclei is much more important. The atomic nuclei in a molecule are not in a static position, but perform an oscillatory motion whereby the centre of mass of the molecule is kept at rest. The atoms (that is, the nuclei with their inner shell electrons) are said to *vibrate* around an equilibrium position, and the frequency of this oscillation is called the *vibrational frequency*. As will be seen later, the number of different vibrational modes increases rapidly with the size of the molecule. For the present discussion it is more important that in each mode the amplitude and hence the energy of the vibration cannot assume any arbitrary value, but is restricted to certain discrete values. In other words, the vibrations of a molecule are quantized.

As an example, let us consider the spectrum of the diatomic molecule CN. The molecule has only one mode of vibration, with the carbon and nitrogen atom oscillating back and forth in opposite directions. If this molecule is irradiated with photons with a wavelength of 388 nm, the photon energy of 25 800 cm^{-1} (3·20 eV) is just sufficient to raise a bonding electron from the ground state to a higher molecular orbit. The amplitude of the atomic oscillation and the vibrational energy of the molecule are left unchanged. If we irradiate the molecule with photons of 359 nm, we have available an energy of 27 800 cm^{-1}. Again 25 800 cm^{-1} is used to raise an electron to a higher energy level and the remaining 2000

* It is worth mentioning, however, that such phenomena as nuclear spin (§ 10.1) and different isotopes (§ 9.1) were first observed in extremely refined studies of atomic spectra.

cm^{-1} is just sufficient to increase the vibrational energy of the molecule stepwise from its lowest value (vibrational ground state) to the next higher value. The emission spectrum of cyanogen radicals in an electrical discharge (Plate 1b) is now readily explained. The strong emission band at 388 nm results from an excited bonding electron receding to the electronic ground state without any change in vibrational energy. The moderately strong emission band at 359 nm results from the combined loss of electronic energy (25800 cm^{-1}) and one unit of vibrational energy (2000 cm^{-1}). The weak emission band at 421 nm results if the molecule loses its electronic energy, but is left in the first excited vibrational energy level; the net energy loss is then

$$25800 - 2000 = 23800 \text{ cm}^{-1}$$

If the molecule were to recede to the second excited vibrational energy level, the net energy loss would be

$$25800 - 4000 = 21800 \text{ cm}^{-1}$$

and we predict a further emission band at 460 nm. This band can indeed be observed experimentally, although it is very weak.

Another example is presented in Fig. 2.3, which shows the absorption spectrum of benzene dissolved in cyclohexane. The energy required to raise a bonding electron of benzene to a higher orbit is 37400 cm^{-1}, which corresponds to a wavelength of 268 nm. However, as a spectral transition this is forbidden by selection rules. Absorption of electromagnetic radiation is allowed if, in addition to the electronic excitation, the vibrational energy of the atomic nuclei increases, which requires multiples of 900 cm^{-1}. This leads to an absorption spectrum with bands at regular intervals of 900 cm^{-1}.*

Molecules *in a gas* have a third possibility of losing or gaining radiation energy. Gaseous molecules can rotate freely around an axis through the centre of mass, but again the rotational motion is quantized. The speed of revolution can only increase by discrete steps and the energy required to increase the speed by one unit is extremely small (about 1 cm^{-1}). As a result, the vibrational bands

* Strictly speaking a second vibrational mode must be taken into account for benzene, but in the present discussion this may be ignored.

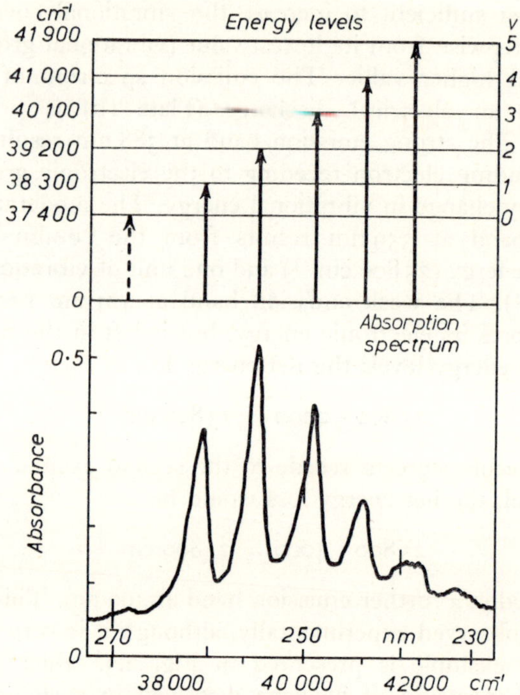

FIG. 2.3. Ultra-violet absorption spectrum of a solution of benzene in cyclohexane. Note the clearly resolved vibration bands. The absorption bands result from transitions from the lowest vibrational level of the electronic ground state to various vibrational levels (v = 1 to 5) of the excited electronic state. This is illustrated in the energy level diagram.

of *gaseous* molecules are split into a great number of separate lines with a wavelength difference less than 0·1 nm. For the cyanogen radical in Plate 1b the wavelength difference between adjacent lines amounts to 0·064 nm. Because of the insufficient resolving power of the spectrometer the lines blend together, which gives rise to the characteristic appearance of molecular band spectra.

In conclusion, the band spectrum of a gaseous molecule consists of a large number of individual *rotation lines*. Each line results from a stepwise change of the potential energy of the molecule that can be expressed as

$$h\nu = \Delta E = \Delta E_{\text{el}} + \Delta E_{\text{vib}} + \Delta E_{\text{rot}} \quad (2.2)$$

where $\Delta E_{el} \gg \Delta E_{vib} \gg \Delta E_{rot}$. ΔE_{el} is the energy required to change the molecular orbit of a bonding electron, ΔE_{vib} is the energy related to a change of the oscillatory amplitude of the atoms in the molecule, and ΔE_{rot} is the energy resulting from a change in rotational speed of the molecule as a whole. The three contributions to the total energy change need not have the same sign.

In solution the molecules no longer rotate freely and the rotational structure of the band spectrum is lost. Thus, the vibrational bands of the dissolved benzene molecule in Fig. 2.3 cannot be resolved into individual rotation lines. Generally, even the vibrational structure of the electronic absorption band is hardly visible and the absorption spectrum of dissolved compounds usually consists of one or two broad, smooth bands.

2.4. Vibration spectra

If in Equation 2.2 ΔE_{el} is put equal to zero, the resulting photon energies are insufficient to excite the electrons in a molecule to higher energy levels, but are still capable of changing the vibrational amplitude of the atoms in the molecule. Such changes are called ***vibrational transitions*** and the corresponding spectrum is called the ***pure vibration spectrum*** of the molecule.

In a diatomic molecule the atoms can only oscillate back and forth along the bonding axis of the molecule. The atoms move in opposite directions and the centre of mass is kept at rest. If the separation between the two atoms is plotted as a function of time, the result is a sinusoidal curve similar to the one shown in Fig. 1.1. This is called a harmonic oscillation, and the molecule is called a *harmonic oscillator*. Quantum mechanical calculations show that the vibrational energy is limited to discrete values and can be expressed as

$$E_{vib} = (v + \tfrac{1}{2})\, h\nu_{vib} \qquad (2.3)$$

where ν_{vib} is the vibrational frequency and v is the *vibrational quantum number* with possible values of 0, 1, 2, ... The energy of the vibrational ground state is equal to $E(v = 0) = \frac{1}{2}\, h\nu_{vib}$. The higher vibrational energy levels are evenly spaced with a separation of $h\nu_{vib}$ (compare the energy level diagram of benzene in Fig. 2.3).

For a molecule consisting of two atoms with masses m_1 and m_2, the vibrational frequency is given by

$$\nu_{vib} = \frac{1}{2\pi}\sqrt{\frac{f(m_1+m_2)}{m_1 m_2}} = \frac{1}{2\pi}\sqrt{\frac{f}{\mu}} \qquad (2.4)$$

where $\mu = m_1 m_2/(m_1+m_2)$ is called the *reduced mass* of the diatomic molecule and f is the force constant that is determined by the electronic forces that hold the atoms together.

For the molecule CO the force constant is equal to 1870 newton/metre and the reduced mass is equal to 6·85 atomic mass units or $1{\cdot}14.10^{-26}$ kg. The vibrational frequency is thus calculated as $6{\cdot}45.10^{13}$ Hz and the energy separation between adjacent vibrational levels is $4{\cdot}26.10^{-20}$ J or 2150 cm^{-1}. A stepwise change of the vibrational energy by one unit can be induced by photons with a frequency of $6{\cdot}45.10^{13}$ Hz, and this corresponds to a wavelength of 4·65 μm. Consequently, the pure vibrational transition of the CO molecule is observed in the infra-red region of the electromagnetic spectrum (Fig. 2.4).

Larger molecules can perform various oscillatory motions. A molecule with N atoms (where obviously N is at least equal to two) has $3N-6$ *different* oscillatory modes (or $3N-5$ for a linear molecule). Each oscillation is characterized by its own vibrational frequency and by a series of vibrational energy levels expressed by Equation 2.3. For each vibration all atoms of the molecule are in oscillatory motion and, accordingly, the vibrational frequencies are determined by the entire molecule and can no longer be expressed by simple formulae. As will be seen in Chapter 5, however, a number of vibrations in larger molecules can be ascribed to the oscillation of only two atoms in the molecule and the corresponding vibrational frequency is described approximately by Equation 2.4.

Transitions between various vibrational levels can be induced by electromagnetic radiation and give rise to absorption bands in the infra-red region of the spectrum. Examples of pure vibration spectra will be considered in Chapter 5. Conversely, the infra-red absorption spectrum of a substance offers the opportunity to determine the vibrational frequencies and the force constants of a molecule. However, there are two main restrictions.

(*a*) In the pure vibration spectrum the selection rule for the absorption of electromagnetic radiation is $\Delta v = +1$. As the separation between two vibrational levels is of the order of 1000 cm^{-1}, virtually all molecules are in the vibrational ground state ($v = 0$) at room temperature (compare Table 1.2). The pure vibration-absorption spectrum then shows only the transitions from the ground state of each vibrational mode to the first vibrational level, the so-called 0–1 transitions or fundamentals.

(*b*) Only those vibrations can be excited by electromagnetic radiation, for which the dipole moment of the molecule varies *during the vibration.* The symmetric vibration of the molecule Br_2 cannot be seen in the infra-red spectrum, because during the oscillatory motion of the bromine atoms the dipole moment remains zero. Oscillations with varying dipole moment, such as the vibration of HCl, are said to be *infra-red active*, since they show up in the infra-red spectrum. Of course, the intensity of different absorption bands can vary appreciably.

2.5. Rotations

The electronic spectrum of the *gaseous* CN-radical in Plate 1b was seen to consist of a multitude of (overlapping) rotational lines. In exactly the same way the infra-red absorption spectrum of the *gaseous* CO molecule consists of separate rotation lines. The quantum mechanical expression for the rotational energy of a diatomic molecule is

$$E_{\text{rot}} = \frac{h^2}{8\pi^2 \mu r_e^2} J(J+1) \tag{2.5}$$

Here μ is the reduced mass of the molecule, r_e is the equilibrium distance between the two atoms and J is the *rotational quantum number*, which can assume the values 0, 1, 2 . . . . For carbon monoxide $r_e = 1{\cdot}128.10^{-10}$ m, and the reduced mass was calculated in the preceding section as $1{\cdot}14.10^{-26}$ kg. With the value for Planck's constant, h, we find

$$E_{\text{rot}} = 3{\cdot}83.10^{-23}\, J(J+1)\ \text{joule} = 1{\cdot}93\, J(J+1)\ \text{cm}^{-1}$$

Fig. 2.4 shows a few rotational energy levels superimposed on the lower two vibrational energy levels of the carbon monoxide molecule. The combined rotation-vibration spectrum of CO can be deduced from this energy level diagram by means of the selection rule that the rotational quantum numbers of the two combining levels must differ by one unit ($\Delta J = +1$ or -1). The wavenumber separation between two successive rotational lines is $3{\cdot}86\ cm^{-1}$.

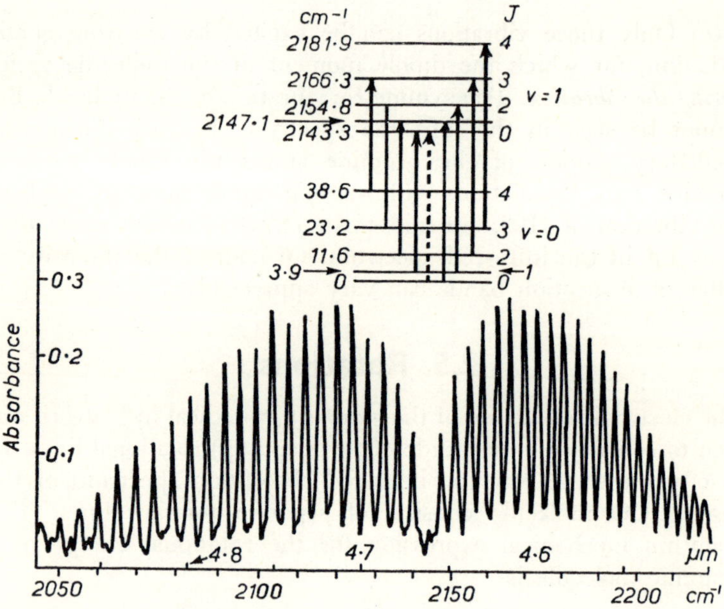

FIG. 2.4. Rotation-vibration absorption band of carbon monoxide. Rotation lines result from transitions between rotational energy levels of the two lowest vibrational levels ($v = 0$ and $v = 1$). Note the absence of the $\Delta J = 0$ transition, which is forbidden by selection rules. Also note that the energy distance between the vibrational energy levels ($J = 0$) has been strongly reduced.

The rotational structure of the pure vibration spectrum can only be obtained for gaseous substances. It is not observed in spectra of molecules in the condensed phase, such as in solution (compare Fig. 5.1).

The *pure* rotation spectrum of a substance is based upon transitions between rotational energy levels without a change of the vibrational or electronic energy. Thus, both ΔE_{el} and ΔE_{vib} in Equation 2.2 are zero. Equation 2.5 predicts that the separation between adjacent rotational levels is about 1 to 10 cm^{-1}. Accordingly, the wavelength of the corresponding spectral lines is of the order of 0·1 to 1 cm and pure rotation spectra are found in the microwave region. Pure rotation spectra can only be obtained for substances possessing a permanent dipole moment, such as CO, but not for CO_2. Also, the substance must be studied at very low pressure (10^{-6} atm.). Although the interest in pure rotation spectra is currently increasing, possible *analytical* applications of rotation spectra are severely restricted by these two limitations and by instrumentation problems.

2.6. Intensity and width of spectral lines

Let us consider a sodium vapour. It is clear from Table 1.2 that even at elevated temperatures virtually all sodium atoms are in the ground state of lowest energy. Now, if the vapour is irradiated with radiation of all possible wavelengths (continuum radiation, § 2.7), the sodium atoms can undergo many absorption transitions allowed by the selection rules (Fig. 2.2). However, if we actually determine the absorption spectrum, we observe only a strong doublet at 589 nm and a much weaker doublet at 330 nm. Other transitions are not observed.

This is due to the fact that the chances of a sodium atom being raised from the ground state to a higher energy level are not equal for all transitions. Each combination of two energy levels in an energy level diagram is characterized by a *transition probability*. If the transition probability is zero, the chance that an atom goes from the one level to the other *under the influence of electromagnetic radiation* is also zero. Between these two levels no spectral transition can take place and the corresponding spectral line cannot be observed in the spectrum. (Note, however, that a *radiationless* transition between the same two levels may still be feasible.)

In principle, all spectral lines with non-zero transition probability should be observable. However, the transition probabilities of various spectral lines can be quite different. For example, the transition probability of the sodium doublet at 589 nm is ten times larger than that of the doublet at 330 nm. A small transition probability yields a comparatively weak spectral line, whereas a large transition probability produces a relatively strong line. This is true for absorption and emission spectra, for atomic as well as for molecular transitions. As will be seen in the expressions derived in Chapters 4 and 6 for the emission intensity and the absorption strength of spectral transitions, the experimentally observed intensities also depend upon instrumental factors, but in all intensity equations the transition probability or a related quantity appears.

The spectra of gaseous atoms or molecules were seen to consist of sharp lines. Whether or not two lines of nearly equal wavelength will be observed separately usually depends upon the resolving power of the spectrometer (Chapter 3). However, a spectral line itself is not purely monochromatic, but extends over a small wavelength region which is called the *spectral line width*. Ultimately, the resolution obtainable in a spectrum is determined by this spectral line width. The line width is also an important factor in atomic absorption spectrometry (§ 6.4).

For gaseous substances the spectral line width in the optical region is predominantly determined by two processes.

(*a*) Generally, a substance must be heated to obtain free molecules or free atoms. Because of the molecular motion resulting from the high temperature the particles have a certain velocity either towards the spectrometer or away from it. This leads to the *Doppler* broadening of absorption and emission lines, which can be expressed as

$$\Delta\lambda_D = \frac{\lambda}{c}\sqrt{\frac{5{\cdot}65\,kT}{M}} \tag{2.6}$$

Here T is the absolute temperature of the gas in degrees Kelvin, k is Boltzmann's constant, and M is the mass of the atom or the molecule (in kg). As an example, the Doppler width of an iron line at 372 nm and at a temperature of 5000°K is equal to 0·0025 nm.

(*b*) The particles in a gas are continuously colliding. Such collisions disturb the process of radiation and lead to *collisional* broadening of the spectral line, which is of the same order of magnitude as the Doppler broadening. The total spectral line width of a gaseous substance (atomic line or rotation line) is thus of the order of 0·005 nm (see problem 7; for yet another cause of line broadening see § 10.2).

In a solution the rotation lines of the molecular spectrum can no longer be observed, because the free rotation of molecules is hindered. The much larger density of solutions strongly enhances the effects of collisions and molecular interaction. As a result, the vibration bands of the pure vibration spectra are strongly broadened and frequently overlap each other. For the same reason the vibration bands of the electronic spectra of dissolved molecules merge together and the vibrational structure of such spectra is only rarely resolved. Thus, the clearly separated vibration bands of the benzene spectrum in Fig. 2.3 are an exception rather than the rule.

2.7. Continuum spectra

Plate 1c presents the spectrum of an ordinary light bulb. Even with a spectrometer of extremely high resolving power no individual lines or bands can be recognized. The light bulb emits radiation of all wavelengths and, accordingly, the spectrum is called a *continuum spectrum.*

Again, radiation of a particular wavelength results from the emission of a photon with a corresponding loss of radiation energy by the glowing wire. However, whereas in free atoms or molecules the photon emission is limited to certain values determined by the separation of the energy levels in the level diagram, the glowing wire can emit photons of any magnitude. Thus the light bulb emits radiation of any wavelength. However, the intensity is not equal for all wavelengths. It is practically zero in the ultra-violet and towards the infra-red region, but reaches a maximum near the visible region (Fig. 2.5). This phenomenon was well known at the end of the nineteenth century, but it could not be explained. In fact, the quantization of radiation energy (the photon concept) was introduced by Planck in 1901 to account for the intensity distribution of continuum spectra. Long-wavelength radiation is readily emitted by the light bulb, but the energy of the individual photons is so small that the total intensity in the infra-red region remains very small. At the other end of the spectrum the photon energy is so large, that the required excitation of electrons in the glowing wire is highly improbable on account of Boltzmann's equation (Equation 1.6).

In this way Planck was able to derive a theoretical expression for the intensity emitted by an ideal lamp (a so-called black body), which is

$$B_\lambda = \frac{2hc^2}{\lambda^5(e^{hc/\lambda kT} - 1)} \tag{2.7}$$

Here B_λ is the *spectral radiance*, which is the radiation energy emitted by the lamp per second, per unit of radiating surface, per

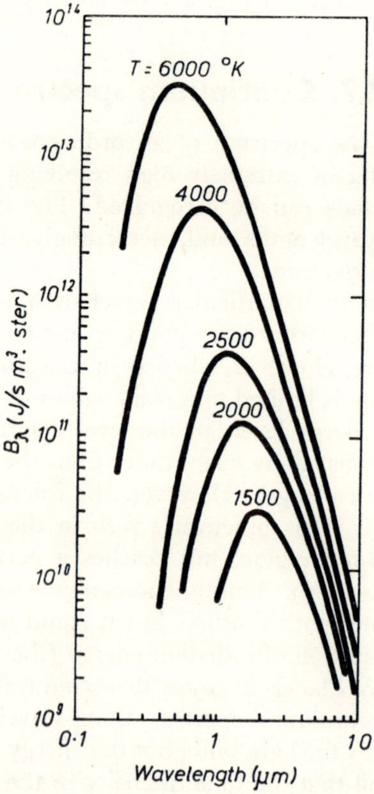

FIG. 2.5. Spectral radiance of a black body at several temperatures, calculated from Equation 2.7.

unit of solid angle* and per unit of wavelength. Fig. 2.5 shows the spectral radiance as a function of the wavelength for black body sources of various temperatures. At each wavelength the intensity increases with temperature and the total amount of radiation emitted ($\int B_\lambda \, d\lambda$) increases with T^4. Also, the maximum of the intensity curve shifts towards smaller wavelengths with increasing temperature: $\lambda_{max} T = 2900 \ \mu m \ °K$.

The surface of the sun has a temperature of about 5000°K and hence the intensity of sunlight is high between 400 and 800 nm. This explains why during their evolution the eyes of man and many animals became sensitive to just this wavelength region; hence we speak of the *visible* region. The glowing wire of a light bulb has a temperature of about 2500°K. Continuum sources of radiation with temperatures between 1500 and 3000°K are widely applied in molecular absorption spectrometry, and Fig. 2.5 will be referred to in Chapters 4 and 6.

Bibliography

1. CANDLER, C., *Atomic Spectra*, Adam Hilger, London, 2nd ed., 1964.
2. HERZBERG, G., *Atomic Spectra and Atomic Structure*, Dover, New York, 1944.
3. KING, G. W., *Spectroscopy and Molecular Structure*, Holt, Rinehart and Winston, New York, 1965.
4. DUNFORD, H. B., *Elements of Diatomic Molecular Spectra*, Addison-Wesley, Reading, Mass., 1968.
5. HERZBERG, G., *Molecular Spectra and Molecular Structure*, Van Nostrand, Princeton, N.J.
   Vol. I. *Spectra of Diatomic Molecules*, 2nd ed., 1950.
   Vol. II. *Infra-red and Raman Spectra of Polyatomic Molecules*, 1945.
   Vol. III. *Electronic Spectra and Electronic Structure of Polyatomic Molecules*, 1966.
6. PEACOCKE, T. A. H., *Atomic Theory and Structure of the Atom*, Pergamon Press, Oxford, 1967.
7. DODD, R. E., *Chemical Spectroscopy*, Elsevier, Amsterdam, 1962.

* A lamp emits radiation in all directions. If we visualize a large sphere with radius r around the lamp, all radiation passes the surface of the sphere. The total solid angle is then the spherical surface divided by the square of the radius, that is $4\pi r^2/r^2 = 4\pi$ *steradians* (see also § 3.7).

Problems

1. From Equation 2.1 deduce an expression for the wavelength of the photon emitted by the electron in the hydrogen atom receding from a level with principal quantum number n_a to a level with principal quantum number n_b. Calculate the wavelength for $n_a = 4$ and $n_b = 3$, and compare the result with Fig. 2.1.

2. How many electrons can be placed in the atomic M-shell?

3. What are the quantum numbers of the outer electron in the aluminium atom?

4. From the wavelengths indicated in Fig. 2.2 calculate the energy difference between the levels $s = \frac{1}{2}$ and $s = -\frac{1}{2}$ for $l = 1$ and $n = 3$.

5. The ultra-violet spectrum of the radical OH is similar to the spectrum of CN in Plate 1b. If a band at 306 nm arises from a purely electronic transition what is the value of ΔE_{el} for OH? Explain the origin of a band at 346 nm. From both bands calculate the vibrational frequency and the force constant of the OH-radical.

6. The pure vibration absorption band of HCl is observed at 2900 cm^{-1} in the infra-red region. Calculate the force constant of the molecule. The force constant will be equal for both modifications, $H^{35}Cl$ and $H^{37}Cl$, but, of course, the reduced masses differ slightly. As a result, the rotation-vibration lines in the absorption band of the gaseous molecule are split into doublets. Calculate the difference in wavenumber between two lines of the same doublet.

7. Calculate the Doppler width of the following spectral lines:

   (*a*) Atomic absorption line of Pb at 283 nm in a flame of 2500°K.
   (*b*) Atomic emission line of Al at 396 nm in an arc discharge of 6000°K.
   (*c*) Rotation-absorption line of CO at 2170 cm^{-1} at room temperature.

8. By differentiating Equation 2.7 show that $\lambda_{max} T = 2900$ μm °K.

CHAPTER THREE

MONOCHROMATORS FOR THE OPTICAL REGION

3.1. Basic principle of a monochromator

If we introduce a little NaCl into the flame of a laboratory burner, we observe the characteristic yellow light emitted by liberated sodium atoms. When we introduce a little LiCl into the flame, we see the red emission of lithium. However, if we introduce a mixture of both compounds into the flame, the colour of the flame depends upon the concentration ratio of the two elements, and the eye is unable to unravel the colour mixture into its two components.

Replacing the eye by an instrumental detection system cannot improve the situation. For the optical region of the electromagnetic spectrum no detectors are known that can separate composite radiation into its components. (As will be discussed in Chapters 8 and 9, such detectors do exist for other parts of the spectrum.) It is therefore necessary to separate the radiation of different wavelengths before it reaches the detector.

In simple cases optical filters may be applied. If a flame with sodium and lithium atoms is viewed first through a yellow filter and next through a red filter, the presence of both elements can be detected *consecutively*. Optical filters are frequently used for a rough separation of wavelengths. However, as the wavelengths approach each other more closely, it becomes increasingly difficult to separate them with filters. Moreover, it is not possible to record a large number of wavelengths simultaneously in this way.

Therefore, another approach is followed to unravel a complicated spectrum. Radiation of different wavelengths is forced to follow different directions, so that the wavelengths are separated *in space*. The spectra shown in Plate 1 consist of *spatially separated images* of the wavelengths emitted by the source of radiation. Here the term image is used in a metaphorical sense. A spectral line represents radiation of a certain wavelength and thus confirms the presence of this radiation just as an image confirms the presence of an object. It is not essential whether this image is formed on a photographic plate (as in Plate 1) or traced onto the chart paper of a recorder (as in Figs. 2.3 and 2.4). In the spectrum each point corresponds to radiation of a single, fixed wavelength, and the appropriate instrument is called a *monochromator* (literally: one-colour-maker).

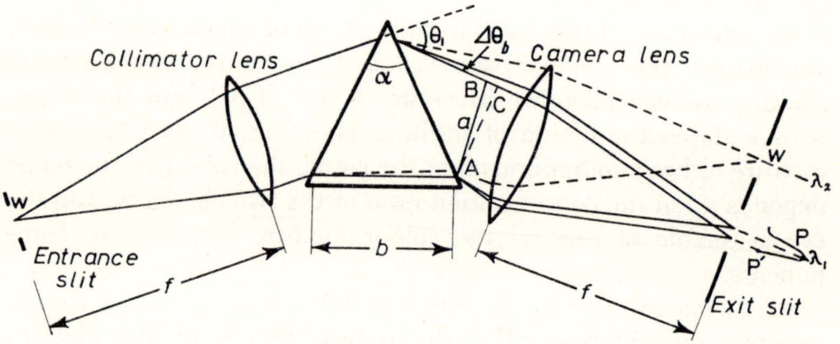

FIG. 3.1. Operating principle of a monochromator.

All monochromators for the optical region (wavelengths between 0·2 and 25 μm) are based upon the principle shown in Fig. 3.1. A polychromatic beam of radiation with various wavelengths strikes a narrow slit. The length of this *entrance slit* (perpendicular to the drawing) is typically a few millimetres, whereas the width may be ten to one hundred times smaller. The entrance slit is placed at the focal point of an achromatic lens (the collimator) which converts the diverging beam into a parallel beam. The beam then meets the central part of the monochromator, a *dispersing device* with the property that radiation of different wavelengths is deflected over different angles. Two dispersing devices considered in the following sections are the prism and the grating.

As a result of this dispersion the original beam of radiation is separated into a number of beams, each of which is a parallel beam containing radiation of a single wavelength and emerging at a different angle θ. With a secondary lens (the camera lens) each beam is focused to give a sharp image of the entrance slit. In the focal plane of the camera lens a series of images results, each one being formed by radiation of a single wavelength. If a photographic plate is placed along the focal plane, the images will be visible after developing as thin black lines (compare the spectrum in Plate 1a). Each spectral line is an *image of the entrance slit* formed by radiation of a single wavelength emitted by the source.

Alternatively, a number of narrow slits may be placed in the focal plane. Each *exit slit* collects radiation of a very small wavelength interval. Finally, it is possible to scan the focal plane with a single exit slit. The line-images in the focal plane are then transmitted one after the other and recorded successively. In practice, it is more convenient to keep the exit slit fixed and rotate the prism or the grating. In that case, the spectrum slowly passes the exit slit and the various monochromatic beams emerge one at a time. These different arrangements will be discussed in Chapters 4, 5 and 7.

3.2. Resolving power of a monochromator

The characteristic quantity of the dispersing device is the *angular dispersion*, $d\theta/d\lambda$, which denotes the *difference* ($d\theta$) in emergent angle corresponding to a difference ($d\lambda$) in wavelength. (Angular differences are measured in *radians*; 1 radian is equal to 57·3°.) As a result of the difference in *deviation* θ there will be a separation between two images of different wavelengths in the focal plane of the camera lens.

The *linear dispersion* is the geometrical distance (ds) corresponding to a difference in wavelength ($d\lambda$). If the focal length of the camera lens is f, the linear dispersion is

$$\frac{ds}{d\lambda} = f\frac{d\theta}{d\lambda} \tag{3.1}$$

Two beams of radiation with wavelengths λ_1 and λ_2 give rise to two images of the entrance slit in the focal plane of the camera lens. The separation between the centres of both images is equal to $(\lambda_1 - \lambda_2)\, ds/d\lambda$. For a good resolution this distance must be more than half the sum of the widths of the images. In most monochromators the focal lengths of the collimator and the camera lens are equal, and hence the magnification is unity. If the width of the entrance slit is w, then the width of each image is also w (unless the width is very small; see below) and two wavelengths will be just resolved when

$$(\lambda_1 - \lambda_2)_{\min} = \Delta\lambda_s = \frac{w}{\mathrm{d}s/\mathrm{d}\lambda} = w\frac{\mathrm{d}\lambda}{\mathrm{d}s} = \frac{w}{f}\cdot\frac{\mathrm{d}\lambda}{\mathrm{d}\theta} \qquad (3.2)$$

$\Delta\lambda_s$ is called the *spectral bandwidth*, and this is given as the product of the slit width w and the *reciprocal linear dispersion* $\mathrm{d}\lambda/\mathrm{d}s$ of the monochromator. If, as is normally the case in analytical spectrometry, the slit width is not too small, *the spectral bandwidth determines the practical resolving power of the monochromator*. For a given monochromator f and $\mathrm{d}\lambda/\mathrm{d}\theta$ are fixed, whereas the slit width w is usually variable. According to Equation 3.2 the spectral bandwidth increases and the practical resolving power decreases with increasing slit width. An example is presented in Fig. 4.6.

It would appear from Equation 3.2 that the spectral bandwidth may be decreased at will by decreasing the slit width or by increasing the focal length of a monochromator. However, this is incorrect, because at very small slit width ($w < 25\ \lambda$, where λ is the wavelength of the radiation) another phenomenon must be taken into account. The prism in Fig. 3.1 not only causes dispersion, but also limits the diameter a of the parallel beam of radiation; thus, it acts as an aperture. As was mentioned in § 1.1 the waves of radiation will not pass the aperture exactly along straight lines, but will spread out to both sides. This is known as diffraction. In Fig. 3.1 it means that radiation of a single wavelength λ_1 will not emerge at an angle θ_1 only, but will also travel in slightly different directions. A beam from one point of the entrance slit and with one wavelength λ_1 will not be focused exactly at one point P, but some radiation will also fall on either side of P. Now, all rays assembling at P have travelled (optically) equal distances. Accordingly, such rays will interfere constructively (§ 1.1) and enhance each other. However, for the beam assembling at P′ the ray through the top of the prism has travelled an additional distance BC as compared to the ray through the prism base. The waves meeting at P′ will now be of different phase and upon interference the intensity will be reduced. If the distance BC is just equal to one wavelength (BC = λ_1), then the ray through the centre of the beam travels an additional distance of $\frac{1}{2}\lambda_1$ and this ray will be in opposite phase to the ray through point A. Consequently, the upper and the lower half of the beam will extinguish each other and the intensity at P′ will be zero.

In this way radiation of wavelength λ_1 and resulting from one point of the entrance slit gives rise to a small area of light with an intensity having a maximum at P and gradually diminishing towards P′. The small angular difference between the beams towards P and P′ can be derived from Fig. 3.1 as

$$\Delta\theta_d = \frac{BC}{a} = \frac{\lambda}{a}$$

and this corresponds to an apparent wavelength difference of

$$\Delta\lambda_d = \frac{d\lambda}{d\theta}\,\Delta\theta_d = \frac{d\lambda}{d\theta}\cdot\frac{\lambda}{a} \qquad (3.3)$$

This may be considered as the theoretical lower limit for the spectral bandwidth. At increasingly smaller slit width the spectral bandwidth $\Delta\lambda_s$ in Equation 3.2 does not approach zero, but is reduced to the diffraction limit $\Delta\lambda_d$. Accordingly, the *theoretical resolving power* of a monochromator is defined as

$$R = \frac{\lambda}{\Delta\lambda_d} = a\,\frac{d\theta}{d\lambda} \qquad (3.4)$$

A numerical example shows that the diffraction broadening of a spectral line is usually negligibly small. For a wavelength of 0·6 μm (yellow light) the angular dispersion of a glass prism is $d\theta/d\lambda$ = 0·06 μm^{-1}. For a beam diameter of 8 cm, the diffraction width is

$$\Delta\lambda_d = \frac{0{\cdot}6}{8.10^4}\cdot\frac{1}{0{\cdot}06}\,\mu m = 0{\cdot}12 \text{ nm}$$

For a focal length of 50 cm the spectral bandwidth is calculated from Equation 3.2 as

$$\Delta\lambda_s = \frac{w}{50.10^4}\cdot\frac{1}{0{\cdot}06} = \frac{w}{3}\,.\,10^{-4}$$

Even at a very small slit width of w = 0·01 mm (= 10^4 nm) the spectral bandwidth is still three times as large as the diffraction width. This confirms our previous remark that in *analytical* practice diffraction broadening is rarely important. The theoretical resolving power expressed by Equation 3.4 is hardly ever reached in analytical spectrometry and it is usually quite correct to calculate the spectral bandwidth from Equation 3.2. Accordingly, manufacturers of monochromators and spectrometers usually quote the reciprocal linear dispersion ($d\lambda/ds$) in the brochures of their instruments.

3.3. The angular dispersion of a prism

In Fig. 3.2 a light ray with wavelength λ strikes the front surface of a prism at an angle i_1. At the air–glass boundary the ray is refracted (§ 1.1) according to

$$\frac{\sin i_1}{\sin r_1} = n(\lambda)$$

where $n(\lambda)$ is the refractive index of glass for the wavelength λ. The ray traverses the solid glass prism and is again refracted at the rear surface, according to

$$\frac{\sin i_2}{\sin r_2} = n(\lambda)$$

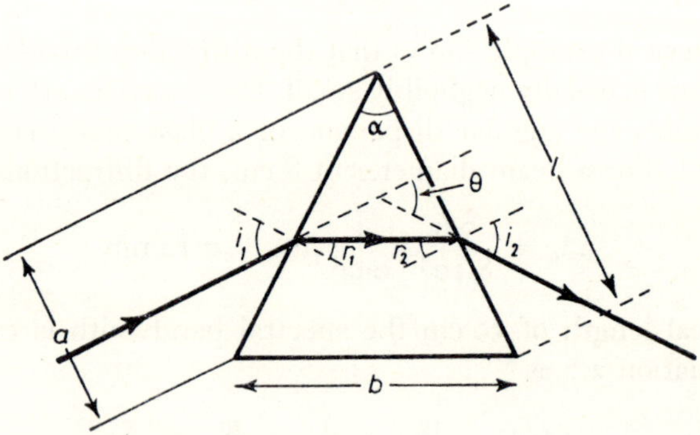

FIG. 3.2. Dispersion of radiation by a prism. A beam with diameter a strikes the prism with refracting angle α, base b and side l, at an angle i_1. The figure shows the ray that traverses the prism at minimum deviation (smallest possible θ).

A second ray with another wavelength (λ'), also striking the prism at an angle i_1, will again be refracted at both boundaries. However, the refractive index $n(\lambda')$ will be slightly different from $n(\lambda)$ and, accordingly, the angles r_1, r_2 and i_2 will take other values. The angle θ between the incident ray and the emerging ray thus depends upon the wavelength of the radiation, and the angular dispersion $d\theta/d\lambda$ of a prism results from the *change of the refractive index with wavelength.*

It can be proved that the angular dispersion is maximal when the angles r_1 and r_2 are equal and hence the angles i_1 and i_2 are equal. Fig. 3.2 then yields the following relations:

$$r_1 = r_2 = \frac{\alpha}{2}$$

$$i_1 = i_2 = \frac{\theta + \alpha}{2}$$

$$n = \frac{\sin i_1}{\sin r_1} = \frac{\sin \dfrac{\theta + \alpha}{2}}{\sin \dfrac{\alpha}{2}}$$

Differentiation yields

$$\frac{dn}{d\lambda} = \frac{1}{2} \frac{\cos \dfrac{\theta + \alpha}{2}}{\sin \dfrac{\alpha}{2}} \cdot \frac{d\theta}{d\lambda} = \frac{1}{2} \frac{\cos i_1}{\sin \dfrac{\alpha}{2}} \cdot \frac{d\theta}{d\lambda}$$

or

$$\frac{d\theta}{d\lambda} = \frac{2 \sin \dfrac{\alpha}{2}}{\cos i_1} \cdot \frac{dn}{d\lambda}$$

Now, if b is the prism base, a is the maximum diameter of the beam and l is the length of the prism edge,

$$\cos i_1 = \frac{a}{l} \quad \text{and} \quad \sin \frac{\alpha}{2} = \frac{\frac{1}{2}b}{l}$$

whence

$$\left(\frac{d\theta}{d\lambda}\right)_{\text{prism}} = \frac{b}{a} \cdot \frac{dn}{d\lambda} \tag{3.5}$$

The angular dispersion of a prism has now been expressed in terms of the characteristic dispersion $dn/d\lambda$ (which is dependent only on the prism material), and the dimensions of the prism. The larger the diffracting angle α of the prism, the larger is the prism base, and the larger is the angular dispersion. In practice, the refracting angle is limited to 60° to keep the reflection losses against the front surface of the prism at an acceptable low level.

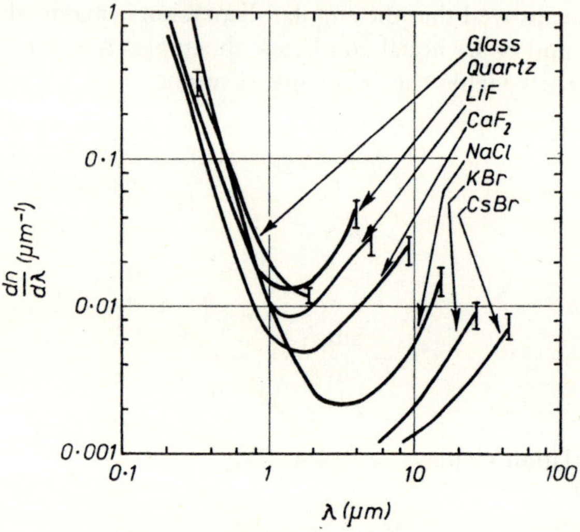

FIG. 3.3. Characteristic dispersion of prism materials for the ultra-violet, the visible and the infra-red regions, as a function of wavelength; vertical lines denote the limit of transparency.

Fig. 3.3 shows the characteristic dispersions of some common prism materials as a function of wavelength. Obviously, the characteristic dispersion varies with wavelength and, consequently, the angular dispersion and the spectral bandwidth of prism monochromators vary with wavelength. Apparently, the characteristic dispersion increases strongly when the wavelength approaches the limit of transparency. In the near infra-red region an LiF prism shows the largest characteristic dispersion and resolving power, but it is applicable up to 5 μm only. On the other hand, sodium chloride is transparent up to 15 μm, which is sufficient for most infra-red applications, but its characteristic dispersion is fairly small, especially at short wavelengths. A similar case occurs in the visible region. Here, the characteristic dispersion of glass is appreciably larger than that of quartz, but the use of quartz is clearly required if spectral investigations are extended into the ultra-violet region.

3.4. The angular dispersion of a grating

Fig. 3.4 shows an extremely enlarged fraction of a *diffraction grating*. Into a piece of highly polished glass, about 10 cm long, a large number of oblique grooves are cut with a small diamond point. The surface is then coated with a thin layer of highly reflective material, such as aluminium for the visible and ultra-violet region. Since it is essential that the spacing d between

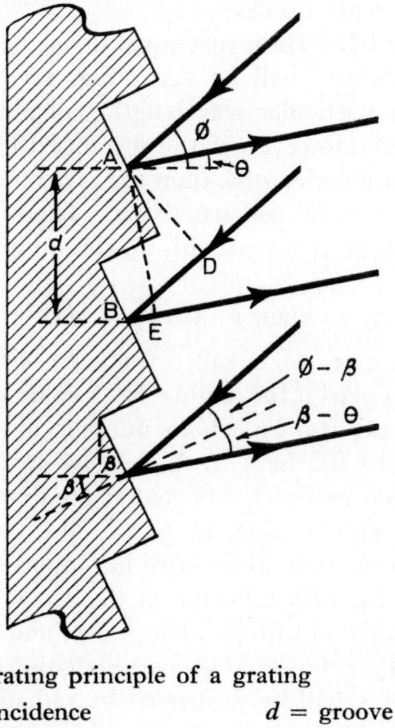

FIG. 3.4. Operating principle of a grating
ϕ = angle of incidence d = groove spacing
θ = angle of reflection β = blaze angle

successive grooves be constant, the ruling must be carried out with extreme accuracy. As a result, such gratings are very expensive. Fortunately, once ruled the glass grating can be used as a master from which plastic imprints can be obtained at much lower cost (*replica gratings*).

The wavelength dispersion of a grating is the combined result of diffraction and interference. Let a parallel beam of radiation with a single wavelength λ strike the grating at an angle ϕ (note that ϕ is measured with respect to the normal to the grating and not with respect to the normal to the groove). Two incident rays of the beam strike the grating at two equivalent points A and B of two adjacent grooves. Radiation is now diffracted into all directions. One such direction is represented by the two rays reflected at an angle θ. The ray reflected at B travels an additional distance

$$BD + BE = d(\sin\phi + \sin\theta)$$

in comparison to the ray reflected at A. If both rays are collected by a lens, interference occurs.

If the distance BD + BE is just equal to one half of a wavelength, the resulting intensity will be zero (Fig. 1.3). This means that radiation of this particular wavelength will not be visible at the emergent angle θ. However, if the distance BD + BE is exactly a whole number of wavelengths, then the rays reflected at successive grooves will interfere constructively. Therefore, radiation of wavelength λ will be reflected with maximum intensity at an angle θ such that

$$d(\sin\phi + \sin\theta) = m\lambda \qquad (3.6)$$

where $m = 0, 1, 2, 3, \ldots$ is called the *order of reflection*.

Now, what happens if the angle θ has a value such that BD + BE is slightly different from a whole number of wavelengths? For example, let BD + BE be $1{\cdot}01\ \lambda$. In that case the path length difference between two rays reflected at the first and the fifty-first groove, respectively, is equal to $50{\cdot}5\ \lambda$ and, consequently, these two rays interfere destructively with net intensity zero. The same argument holds for rays reflected at the second and fifty-second groove, at the third and fifty-third groove, and so on. Therefore, in a grating with one hundred grooves radiation reflected from the upper fifty grooves will be destroyed by radiation reflected from the bottom fifty grooves and no light will be detected at an angle θ

such that the optical path length difference between successive grooves is 1·01 λ. Now, actually, a grating consists of many thousands of grooves and, as a result, for a *particular* wavelength λ the reflection angle θ is very precisely defined indeed.

If the incident beam contains radiation of *various* wavelengths, Equation 3.6 shows that each wavelength is reflected with maximum intensity at a different angle θ. Differentiating Equation 3.6, we obtain for the angular dispersion of a grating

$$\left(\frac{d\theta}{d\lambda}\right)_{\text{grating}} = \frac{m}{d \cos \theta} \tag{3·7}$$

The angular dispersion of a grating does not depend upon the material used for the grating or upon the wavelength. (It is true that the angle of reflection θ depends upon the wavelength (Equation 3.6), but over the spectral range of interest the variation of $\cos \theta$ with λ is usually small.) Accordingly, a grating monochromator shows equal angular dispersion and equal spectral bandwidth over the entire spectrum (see however, § 4.2). According to Equation 3.7 the angular dispersion of a grating increases with decreasing groove spacing d. However, this increase is not unlimited. Since in Equation 3.6 $\sin \phi$ and $\sin \theta$ cannot exceed a value of 1, the spacing d must be greater than 0·5 λ for $m = 1$. For $d < 0{\cdot}5\ \lambda$ the grating loses its dispersive power and acts as an ordinary plane mirror. Consequently, the groove spacing must increase with wavelength. A grating with 1000 grooves per millimetre* ($d = 1\ \mu$m) can still be used in the visible region, but in the infra-red region the spacing must be increased to about 20 μm (see § 5.2).

3.5. Overlapping orders and blaze angle

The angular dispersion of a grating also increases with the *order of reflection*. What does this factor mean? According to Equation 3.6 radiation of wavelength λ is not reflected at a single angle θ only, but at several angles corresponding to $m = 0, 1, 2, 3, \ldots$. In Fig. 3.5 a ray of wavelength 600 nm strikes a grating with a groove

* The common expression is 1000 *lines* per mm (or 25 000 lines per inch), but in order to avoid confusion with *spectral line* the term *groove* will be used.

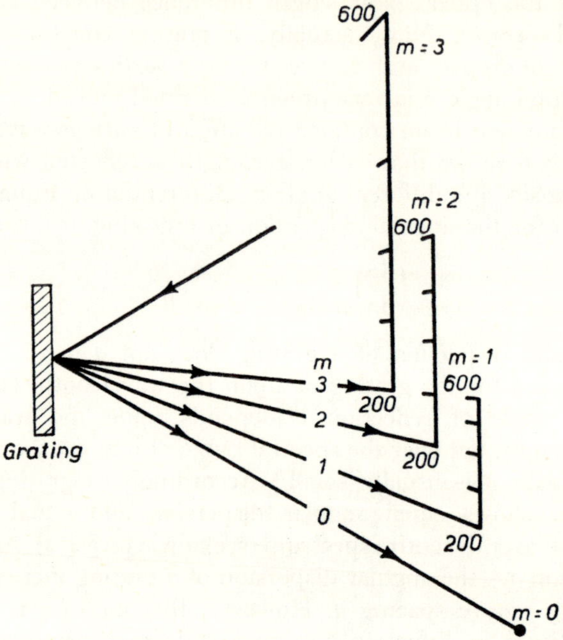

FIG. 3.5. Overlapping orders in grating spectra. A beam of polychromatic radiation with wavelengths from 200 to 600 nm strikes a grating with groove spacing of 1·5 μm at an angle of 30°. The figure shows the length of the spectra reflected in the zeroth, first, second and third orders, and the reflected rays for a wavelength of 200 nm. (Compare Table 3.1.)

spacing of 1·5 μm at an angle $\phi = 30°$. The angles of reflection calculated from Equation 3.6 are shown in the figure and presented in Table 3.1. When the angles of reflection for some other wavelengths are calculated, the following conclusions can be drawn.

(*a*) In the 'zeroth' order ($m = 0$, $\theta = -\phi$*) all wavelengths coincide; no dispersion occurs.

(*b*) With increasing order the spectra cover an increasingly

* A negative angle θ means that the ray is reflected to the other side of the grating normal.

larger distance; this agrees with the increase of the angular dispersion with m (Equation 3.7).

(*c*) Spectra reflected in different orders partially overlap. In Fig. 3.5 the direction of reflection is equal for 600 nm in the first order, for 300 nm in the second order and for 200 nm in the third order. In general, if $\lambda_1 = 2\lambda_2 = 3\lambda_3 = \ldots = m\lambda_m$, then the wavelengths $\lambda_1, \lambda_2, \lambda_3, \ldots \lambda_m$ coincide. If a grating monochromator is used to record a long spectrum, such as an infra-red spectrum from 3 to 15 μm, overlapping of orders must be prevented. This may be done by filters, or, at higher cost, by a small prism. If in Fig. 3.5 a glass filter is placed in the path of the ray reflected at $-6°$, all radiation with wavelength smaller than 350 nm will be absorbed (Fig. 3.3) and only radiation of 600 nm will be detected.

TABLE 3.1. Angle of reflection θ for some wavelengths, striking a grating with groove spacing d = 1·5 μm at an angle of incidence ϕ = 30°

	Angle of reflection θ		
Order m	λ = 600 *nm*	300 *nm*	200 *nm*
0	−30°	−30°	−30°
1	−6°	−18°	−22°
2	+18°	−6°	−13°
3	+45°	+6°	−6°
4	—	+18°	+2°

The reflection of several orders has yet another consequence. In principle, the total incident intensity will be distributed equally over all orders and the intensity of the spectrum reflected in one particular order will be rather low. This disadvantage is greatly reduced, if the grooves are ruled at a certain *blaze angle* β. The incident beam of radiation is then preferentially reflected in the direction that is symmetrical to the normal to the oblique groove. In Fig. 3.4 the incident ray strikes the grating at an angle ϕ to the grating normal and thus at an angle $\phi-\beta$ to the groove normal. In the same way the reflected ray lies at an angle $\beta-\theta$ to the groove normal (recall that θ may be negative!). Now, the reflection will be most intense if

$$\phi-\beta = \beta-\theta \quad \text{or} \quad \theta = 2\beta-\phi$$

For example, if the grooves in the grating of Fig. 3.5 are ruled at a blaze angle $\beta = 12°$, we calculate with $\phi = 30°$ a preferential angle of reflection of $\theta = -6°$. Consequently, radiation with a wavelength of 600 nm will be predominantly reflected in the first order. In practice a first order efficiency of 70 per cent is considered to be quite good.

Two remarks must be made, however. In the first place the grating equation 3.6 must always be obeyed and, consequently, the action of the blaze will only be perfect for one single wavelength in the first order. This is called the *blaze wavelength*. In practice, the blaze action extends over a fairly broad region and a grating blazed for a (first order) wavelength of 600 nm will reflect a high-intensity first order spectrum ranging from 400 to 900 nm. Obviously, however, blazing reduces the dynamic range of the grating. In the second place, blazing does not prevent order-overlap. A grating blazed for 600 nm in the first order will reflect a radiation of 300 nm predominantly in the second order. In other words a strong first order spectrum from 400 to 900 nm is still overlapped by a strong second order spectrum from 200 to 450 nm.

3.6. Comparison between grating and prism

The following points must be considered in the choice between a grating and a prism.

(*a*) In order to reach the intensity of the single prism spectrum the grating must be blazed, but this reduces the dynamic range of the grating.

(*b*) If a spectrum is to be recorded over an extended range, several gratings blazed at different angles must be used alternatively, which increases the costs. On the other hand, it is then possible to utilize for each part of the spectrum a grating with optimum groove spacing, providing an appreciable gain in dispersion (§ 5.2).

(*c*) In a grating monochromator overlapping of orders must be prevented by additional provisions that raise the price of the instrument.

(*d*) Whereas the dispersion of a prism varies with wavelength, the dispersion of a grating is equal over the entire spectrum. When a fixed entrance slit is used, as in emission spectrometry (Chapter 7), the grating monochromator offers a constant resolving power over the entire spectrum (Equation 3.2).

(*e*) No grating is perfect, and unwanted reflections do occur. Consequently, when a grating monochromator is set for a particular wavelength, some radiation outside the spectral bandwidth will emerge from the exit slit. In a prism monochromator the amount of *stray light* is much smaller.

(*f*) The greatest limitation to the use of prisms is the requirement of transparency. Some potentially good prism materials are simply inapplicable, because they can only be used over a very small wavelength range. Other materials may be difficult to shape or sensitive to humidity or temperature influences. These objections do not apply to gratings. Whereas for some spectral regions suitable prisms are simply nonexistent (notably the far infra-red range), good gratings can be manufactured for any spectral region.

(*g*) No doubt the biggest advantage of a grating is its potentially large dispersion, especially when used in higher orders. The advantage of a large dispersion is the more important, since Equation 3.2 shows that for a given spectral bandwidth the grating monochromator allows a larger slit width. As will be seen in the next section, a larger slit width means that more radiation reaches the detector, which improves the sensitivity of the determination.

Summing up, and generalizing somewhat, we may conclude that prism monochromators are relatively cheap and perfectly suitable for simple routine instruments, notably in the visible and near ultra-violet region of the spectrum. A grating requires additional provisions to reduce stray light and to separate the orders, which makes the grating monochromator more expensive. However, if we increase our demands (for example if we want a better resolving power or an extended wavelength range), the use of a grating monochromator becomes imperative. With the cheaper replica gratings now available, the application of grating monochromators has expanded rapidly, especially in the infra-red and the far ultra-violet region of the spectrum.

The larger dispersion obtainable with a grating may be illustrated by a simple example, which will also serve to demonstrate the application of the equations derived above. Let the normal infra-red spectrum between 3 and 15 μm be scanned with an NaCl prism with a refracting angle α of 60° and a base of 10 cm. At 8 μm the refractive index is 1·5 and the characteristic dispersion is equal to 0·005 μm^{-1} (Fig. 3.3). What is the angular dispersion? In Fig. 3.2 $r_1 = \frac{1}{2}\alpha = 30°$ and $\sin i_1 = n \sin r_1 = 0{\cdot}75$; the beam diameter is $a = l \cos i_1 = 6{\cdot}6$ cm. The angular dispersion is thus 0·0075 μm^{-1}. Now, let the same spectrum be scanned by a grating with 50 grooves per mm ($d = 20$ μm). For $\cos\theta \simeq 1$ the angular dispersion is then calculated from Equation 3.7 as 0·05 μm^{-1} in the first order. With respect to the prism the gain is a factor of seven. The advantage of the grating is even larger at shorter wavelengths. Not only does the characteristic dispersion of the NaCl prism decrease slightly, but also the grating may be used in a higher order. Naturally, a good order separation is essential.

In order to record the spectrum between 7 and 15 μm at optimum intensity the blaze wavelength should be about 10 μm. What is the blaze angle? Since it has been assumed that $\cos\theta = 1$, θ must be 0°. Equation 3.6 then yields, with $d = 20$ μm, $\lambda = 10$ μm and $m = 1$, an angle of incidence $\phi = 30°$. With the condition $\phi - \beta = \beta - \theta$ we find that the blaze angle must be 15°. The angle of reflection θ is equal to $-9°$ for a wavelength of 7 μm, 0° for a wavelength of 10 μm and $+15°$ for a wavelength of 15 μm. With the same grating the wavelength range from 3 to 7 μm may be scanned in the second order with an optimum reflectivity at 5 μm (second order angle of reflection $\theta = 0°$).

3.7. The luminosity of a monochromator

The primary requirement of a monochromator is good wavelength separation. However, an important secondary requirement is its capability to gather as much radiation from the source as possible, to increase the sensitivity of the determination or to reduce the task of the detection system behind the monochromator. Unfortunately, these are two conflicting requirements. This may be illustrated as follows. If diffraction effects are momentarily ignored, Equation 3.2 predicts that the spectral bandwidth decreases with decreasing slit width, and at zero slit width the resolving power would be infinitely high. With the slits closed, however, no radiation at all enters the monochromator and instead of recording a highly resolved spectrum we record no spectrum at all.

To discuss this aspect of light gathering power or *luminosity* in more detail, let us consider a molecular absorption spectrometer where a continuum source of radiation is placed in front of a monochromator (Figs. 4.1, 4.2 and Plate 2). Fig. 2.5 shows that the decisive quantity of the source is the spectral radiance expressed in watts per unit of surface, per unit of solid angle and per unit of wavelength. On the other hand, the photoelectric detector responds to the radiative power in watts emerging from the exit slit of the monochromator. Consequently, the monochromator introduces a proportionality factor with the dimensions of a solid angle times a wavelength interval times a surface area, and this factor should be as large as possible.

The solid angle subtended by the monochromator is given by A/f^2, where A is the cross-section of the collimator lens (or the grating surface if this is smaller) and f is the focal length of the collimator. Consequently, with decreasing focal length the luminosity increases, but simultaneously the spectral bandwidth increases (Equation 3.2).

The wavelength interval is given by the spectral bandwidth of Equation 3.2. Consequently, the luminosity decreases with increasing dispersion, again in contrast to the resolving power. Also, the luminosity increases with the slit width.

The surface area, finally, is determined by the dimensions of the entrance slit, which is normally chosen to be equal to the exit slit. The slit height is usually limited to a few millimetres by the dimensions of the source, whereas the slit width is again determined by the balance between high luminosity on the one hand and high resolving power on the other. It is worth noting, however, that the slit width appears both in the spectral bandwidth and in the slit area. Consequently, the amount of radiation transmitted by a monochromator is proportional to the *square* of the slit width in the case of a continuum source. Therefore, in molecular absorption spectrometry, slit width variation is a simple means of regulating the amount of radiative power reaching the detector (§§ 4.2 and 5.2).

Bibliography

1. JENKINS, F. A., and WHITE, H. E., *Fundamentals of Optics*, McGraw-Hill, New York, 3rd ed., 1957.
2. SAWYER, R. A., *Experimental Spectroscopy*, Dover, New York, 1963.
3. TOLANSKY, S., *Revolution in Optics*, Penguin Books, London, 1968.
4. MARTIN, A. E., *Infra-red Instrumentation and Techniques*, Elsevier, Amsterdam, 1966.
5. JAMES, J. F., and STERNBERG, R. S., *The Design of Optical Spectrometers*, Chapman and Hall, London, 1969.

Problems

1. Calculate the spectral bandwidth of monochromators with an entrance slit width of 0·1 mm and reciprocal linear dispersions of 10 nm/mm, 2 nm/mm and 0·5 nm/mm respectively; calculate the corresponding angular dispersions for a focal length of 0·75 m.

2. The following values for refractive indices can be found in the literature.

λ (*nm*)	*NaCl*	*LiCl*	CaF_2
2·357	1·526	1·475	1·421
3·536	1·523	1·473	1·414

prisms cut from these materials, with a base of 10 cm, are placed in a monochromator with a focal length of 50 cm and a slit width of 0·1 mm. The prisms are traversed by a beam of radiation of 3 μm wavelength and 5 cm diameter. Calculate the characteristic dispersion, the angular dispersion and the spectral bandwidth for all three prisms.

3. (*a*) Radiation with a wavelength of exactly 500 nm enters a monochromator with an entrance slit 0·8 mm wide, unit magnification and a reciprocal linear dispersion of 2·5 nm/mm. How wide is the image at the focal plane, and what is the corresponding spectral bandwidth?

(*b*) The focal plane is slowly scanned with an exit slit equal in width to the entrance slit. An attached wavelength counter accurately indicates the wavelength of any radiation that emerges from the centre of the exit slit. What wavelength value is indicated by the counter when radiation is first observed through the exit slit?

(*c*) At which counter value is the exit slit fully illuminated?

(*d*) Draw a graph showing the amount of radiation passing the exit slit as a function of the number read from the wavelength counter.

4. A grating is ruled with 1200 grooves per mm. The angle of incidence is 0°. Calculate the first order angle of reflection for wavelengths of 300, 500 and 800 nm, and the corresponding blaze angles to reflect these wavelengths with maximum intensity in the first order.

5. What infra-red wavelengths between 2 and 15 μm overlap at 15 μm in a grating spectrum?

6. A continuum source of radiation with a temperature of 2500°K is placed in front of a grating monochromator to record an absorption spectrum from 400 to 1000 nm. The increase of the incident intensity between 400 and 1000 nm can be read from Fig. 2.5. In scanning the spectrum the widths of the equal entrance and exit slits are continuously adjusted in such a way that the radiative power reaching the detector is the same at all wavelengths. The slit width is 0·1 mm at 400 nm wavelengths. Prepare a rough graph of the slit width as a function of wavelength.

7. Monochromator A is similar to monochromator B in all respects, except that its linear dispersion is four times larger.

(*a*) Which monochromator has the larger spectral bandwidth?

(*b*) If both monochromators are illuminated with a continuum source of radiation, which one transmits more radiation?

(*c*) If both monochromators should transmit equal amounts of radiation, which one should have a larger slit width?

(*d*) Which monochromator now has the larger spectral bandwidth? What is the difference as compared to question (*a*)?

CHAPTER FOUR

MOLECULAR ABSORPTION SPECTROMETRY

ULTRA-VIOLET AND VISIBLE

4.1. Single-beam spectrometers

Fig. 4.1 shows the optical diagram of a single-beam spectrometer for measuring the absorption of molecular radiation in the ultra-violet and visible region of the spectrum (0·2–1 μm). Radiation from either one of two continuum sources (W or H, selected by the position of mirror M1) is reflected by the concave mirror M3 onto a quartz prism, the rear surface of which is coated with a thin layer of highly reflective aluminium. The prism is thus traversed twice and the refracting angle is effectively doubled to 60°. (This is called a Littrow mounting.) The concave mirror M3 serves as both the collimator 'lens' and the camera 'lens', which results in a compact arrangement. Subsequently, the beam of radiation passes the exit slit S2, traverses the absorption cell and is deflected by mirror M5 to either one of two detectors.

The widths of the entrance slit and the exit slit can be varied simultaneously between 0 and 2 mm. By rotating the prism around an axis perpendicular to the diagram the central wavelength of the radiation emerging from the exit slit can be varied; this wavelength is read from scale G. The detector converts the radiation into an electrical current (§ 4.3) which is amplified and measured with a galvanometer.

The absorption of radiation by a solution is measured in the following way. The appropriate detector and source are chosen, and radiation of a suitable wavelength is selected by rotating the prism. An absorption cell containing pure solvent is placed into the cell compartment and the slit width and the amplification are adjusted to bring the galvanometer to a full scale deflection of 100 scale units. Next, the absorption cell is filled with the sample

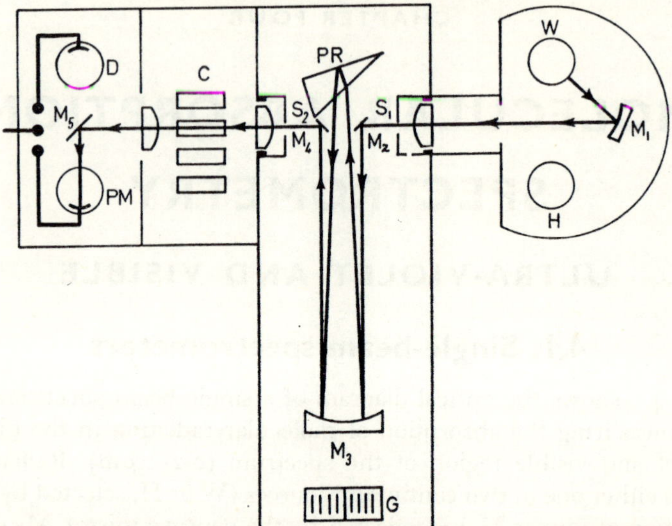

FIG. 4.1. Optical arrangement of a single-beam absorption spectrometer for the visible and ultra-violet spectrum (model PMQ II from Zeiss, Western Germany).

M_1, M_2, M_3, M_4, M_5 = mirrors	S_1 = entrance slit	C = absorption cell
W = tungsten lamp	S_2 = exit slit	PM = photomultiplier
H = hydrogen lamp	PR = prism	D = photoconductor
	G = wavelength scale	

solution and brought into the light path. Because the dissolved sample partly absorbs the radiation, the detector receives a smaller intensity and the deflection of the galvanometer will be diminished, for example to a value of 40 scale units. The sample solution then shows a *percentage absorption* of 60 per cent or a *transmittance* of 40 per cent. An expression relating the transmittance of the solution to the concentration of the absorbing component will be derived in § 4.5.

With respect to this method the following comments can be made.

(*a*) Because some compounds decompose under the influence of ultra-violet radiation, the absorption cells are placed *behind* the monochromator. In this way only a very small fraction of the total radiation emitted by the source falls on the absorption cell.

(*b*) For measurements according to the *deflection method*

described above it is essential that the detection be linear, in order that the reduction of the intensity be accurately converted into a proportional decrease of the galvanometer deflection. Detection need not be linear, if the principle of the 'optical-null' is applied, as will be discussed in § 5.2.

(*c*) With pure solvent in the absorption cell sufficient radiation should reach the detector to permit a full-scale deflection of the galvanometer. In addition to the properties of the detector and the source of radiation (§ 4.3), this is controlled by the amplification and the slit width. If the solvent significantly absorbs radiation, the amplification or the slit width must be increased, which enhances either the noise inherent in the signal or the spectral bandwidth (§ 3.2). For this reason solvents are preferred which show negligible absorption between 0·2 and 1 μm, such as water, ethanol, cyclohexane, etc.

(*d*) Single-beam instruments are perfectly suitable for the quantitative measurement of absorption at a fixed wavelength. In principle, the procedure outlined above can be repeated for consecutive wavelengths and the results can be plotted on graph paper to obtain a complete absorption spectrum. However, this is tedious and time-consuming work, because for each wavelength the slit width or the amplification must be readjusted in consequence of the variation of source intensity and detector sensitivity with wavelength.

(*e*) The spectrometer shown in Fig. 4.1 can be set to any wavelength between 0·2 and 1 μm. In cheaper instruments the monochromator is replaced by a set of interchangeable filters. Such instruments are commonly referred to as *colorimeters*, but this name is also, and more appropriately, used for instruments designed specifically for the measurement of colour; the correct name is *filter photometer*. The width of the spectral band transmitted by a filter (equivalent to the spectral bandwidth of a monochromator) is about 20 nm. Consequently, the resolving power of these instruments is rather limited, but the greater intensity transmitted enhances the precision of the measurement. Filter photometers are used for the quantitative analysis of simple solutions in the visible region.

4.2. Double-beam spectrometers

The automatic recording of an absorption spectrum is feasible with double-beam spectrometers, an example of which is shown in Fig. 4.2. Radiation from either a tungsten filament lamp or a hydrogen discharge lamp enters the entrance slit S_1, is reflected by a concave mirror and dispersed by the grating R, returns to the concave mirror and emerges from the exit slit S_2. (This is called an Ebert–Fastie mounting.)

Radiation of a particular wavelength now falls upon a fast-revolving disk A, one half of which is reflecting, and the other half open. When the beam strikes the mirror section, it traverses the

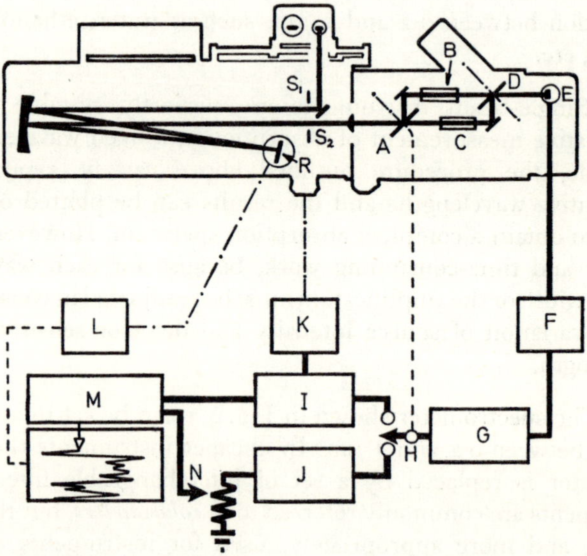

FIG. 4.2. Double-beam spectrometer for the visible and ultra-violet spectrum (model CF4DR, by courtesy of Optica-Milano, Italy).

S_1 = entrance slit
S_2 = exit slit
R = grating, 600 lines/mm
A, D = revolving disks
B = reference cell
C = sample cell
E = photomultiplier
F, G = amplifiers
H = switch
K, L = driving motors
M = recorder

reference cell B and passes through the open section of a second, synchronously rotating disk D. When the beam passes through the open section of disk A, it traverses the sample cell C and is deflected by the mirror portion of D on to the detector E. In this way the detector receives radiation through the reference cell or the sample cell alternately.

The electrical signal of the detector is amplified in F and G. Switch H is connected to the rotating disks A and D. With the mirror section of A in the light path the signal from the reference cell is directed to I by the appropriate position of switch H; the signal from the sample cell is fed into J. If the sample partly absorbs the radiation, the signals from I and J are unequal, and their ratio (§ 4.5) is indicated by the recorder M.

While the recorder chart moves, the motor L slowly turns the grating R, which produces a continuous variation of the wavelength of the radiation emerging from the monochromator. In this way each point on the chart corresponds to a definite wavelength and a complete absorption spectrum is recorded. If the absorption cells B and C are both filled with solvent, the signals from I and J should obviously be equal. Precise adjustments may be made with the potentiometer N.

The intensity of the source and the sensitivity of the detector vary with wavelength. In order to prevent detector fatigue, and the attendant nonlinear response, the electrical signal from the reference cell is continuously compared in I with a constant standard voltage. If the reference signal exceeds the standard voltage, the widths of slits S_1 and S_2 are reduced simultaneously by a driving motor K, and the intensity of the beam of radiation is decreased. When the reference intensity becomes too low, the slits are automatically widened. This principle of '*constant-energy detection*' is utilized in many double-beam spectrometers. As a result of this varying slit width the spectral bandwidth of the monochromator varies with wavelength, even if a grating is applied.

In general, double-beam spectrometers do not yield the quantitative precision obtainable with a single-beam spectrometer. The optics of the monochromator and the quality of the electrical parts determine the quality of the recorded absorption spectrum and the price of the instrument.

4.3. Sources of radiation, detectors and absorption cells

High stability is an essential requirement for the source of radiation, especially in a single-beam spectrometer. Indeed, the source intensity must remain constant during the time required to replace the absorption cell with the reference solution by the cell with the sample solution. The following sources of radiation are commonly applied.

(*a*) The *tungsten filament lamp* closely resembles an ordinary light bulb. A strip or wire of high purity tungsten (inside a silica or glass envelope containing argon to prevent oxidation of the tungsten wire) is heated by an electrical current of about 8 A to a temperature of about 2500°K. According to Fig. 2.5 the emission intensity is maximal in the near infra-red and decreases rapidly towards the ultra-violet, rendering the source unsuitable for wavelengths smaller than 350 nm.

(*b*) The *hydrogen discharge lamp* consists of a silica envelope containing hydrogen gas at a pressure of 0·01 atm. The hydrogen molecules emit a moderately intense continuum spectrum extending from 400 nm down to 190 nm. In addition the lamp emits characteristic atomic hydrogen lines in the visible region which are useful in checking the wavelength scale of the spectrometer. The lamp intensity may be increased by replacing hydrogen by deuterium.

(*c*) The *xenon arc lamp* is essentially an electrical arc discharge (§ 7.1) in xenon gas at 10 atm. The continuum spectrum is extremely intense, but with respect to stability and life-time this lamp is inferior to the preceding two sources. The xenon arc lamp finds its main application in fluorimetry (§ 4.8).

(*d*) The *mercury discharge lamp* emits the line spectrum of mercury vapour excited by an electrical discharge. Hence, this source is not suitable to measure an absorption spectrum, but it can be used for the quantitative measurement of absorption or fluorescence at the fixed wavelength of one of the mercury lines.

The *photomultiplier tube* is the preferred detector for absorption spectrometry in the ultra-violet and visible region of the spectrum. In Fig. 4.3 radiation enters an evacuated glass envelope through a silica window and strikes a *photocathode* coated with a semi-conducting compound, for example caesium antimonide. Incident radiation produces emission of electrons from the photocathode. This requires a minimum energy equal to the binding energy of the electron, which may amount to 2 eV. This corresponds to a photon

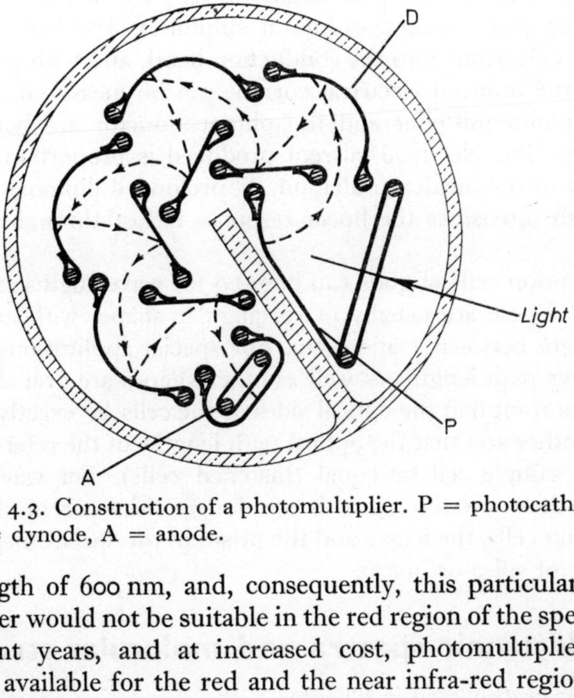

FIG. 4.3. Construction of a photomultiplier. P = photocathode, D = dynode, A = anode.

wavelength of 600 nm, and, consequently, this particular photomultiplier would not be suitable in the red region of the spectrum.* In recent years, and at increased cost, photomultipliers have become available for the red and the near infra-red region of the spectrum.

The electrons emitted by the photocathode are attracted by the first *dynode*, which is kept at a positive potential of 50–100 V with respect to the cathode. The vigorous impact of the electrons

* The fact that even an extremely intense beam of radiation with a wavelength of 800 nm produces no electrons from this photocathode, clearly demonstrates that the photoelectric effect is due to the energy of individual photons. This constitutes convincing evidence for the existence of photons.

liberates an increased number of secondary electrons from the dynode, which are accelerated towards the following, more positive dynode. By combination of ten dynodes the primary cathode current is amplified some 10^6 times and the sensitivity of this detector is extremely large. The sensitivity of the photomultiplier can be changed by varying the total potential drop between the last dynode (the anode) and the photocathode.

For wavelengths larger than 600 nm a cheap alternative to the photomultiplier is the *photoconductor cell*, consisting of a semiconductor (for example cadmium sulphide). Incident radiation promotes electrons into the conduction band, and with an applied voltage the induced electrical current can be measured.

The photomultiplier and the photoconductor are both *linear* detectors. The electrical current produced is proportional to the intensity of the incident radiation. At prolonged illumination with very high intensities the linear response is lost through detector fatigue.

Absorption cells of glass can be used for wavelengths exceeding 350 nm. These are usually of rectangular shape, with an *optical path length* between 1 and 4 cm. For special applications smaller and larger path lengths as well as other shapes are available. It is very important that the optical sides of the cells be exactly parallel to each other and that the optical path lengths of the reference cell and the sample cell be equal (matched cells). For wavelengths smaller than 350 nm glass is no longer transparent, and the absorption cells, the lenses and the prism in the spectrometer must be made of silica or quartz.

4.4. Electronic spectra and molecular structure

As has been discussed in Chapter 2, the absorption spectra in the ultra-violet and the visible region of the spectrum result from transitions of the bonding electrons in molecules. Simultaneously, the vibrational frequency of the atoms in the molecule may be changed, but in the spectra of dissolved compounds the vibrational structure is rarely resolved. This is due to the broadening of vibration bands, to the great number of these bands in larger molecules, and to the interaction between the dissolved molecules, either mutually or with solvent molecules.

Accordingly, the absorption spectra in this part of the electromagnetic spectrum somewhat resemble a landscape with broad hills and valleys, without sharp peaks or other striking features. The spectra of similar compounds largely coincide and, therefore, the electronic spectrum of a molecule is much less suitable for qualitative identification purposes than the vibrational spectrum in the infra-red region (Chapter 5).

However, some correlation between the molecular structure of a compound and its electronic absorption spectrum is unmistakable, and, taken with other data (infra-red spectrum, nuclear magnetic resonance spectrum), the ultra-violet spectrum can provide valuable information in qualitative organic analysis. Notably, the strength of the absorption (the absorptivity, § 4.5) is an important aid. It is also helpful in quantitative and inorganic analysis to know if and where a particular substance may be expected to absorb

TABLE 4.1. Absorption bands in the visible and ultra-violet spectrum

Substance	*Wavelength (nm)*	*Absorptivity (l/mol cm)*	*Solution colour*
Benzene	254	204	—
Toluene	261	225	—
Nitrobenzene	269	7800	—
Chlorobenzene	264	190	—
Aniline	288	1900	—
Aniline	234	9100	—
Naphthalene	275	316	—
Anthracene	375	630	Faintly yellow
Phenanthrene	296	12 500	—
Tetracene	474	—	Orange
Pentacene	580	12 500	Blue
Acetate	220	45	—
Nitrate	301	63	—
Nitrite	354	29	—
Cupric	800	11	Blue
Ferric	335	220	Yellow
Permanganate	525	2400	Pink
Manganate	603	1600	Green
Dichromate	455	420	Orange
Chromate	374	4800	Yellow
Cobaltous	610	50	Green
Cobaltic	510	7	Pink

radiation in the region between 0·2 and 1 μm. Therefore, some basic rules will be briefly discussed.

Rather a lot of energy must be supplied to raise an electron in a *single* bond to a higher energy level. The absorption bands of saturated hydrocarbons, water and simple salts (NaCl, KBr) are found at 100–200 nm, which is outside the region normally considered in molecular absorption spectrometry. *Double* bonds are more easily excited, and such gases as O_2 and N_2 absorb even at 180–200 nm. Consequently, air must be removed from the spectrometer for the study of wavelengths smaller than 200 nm, and this is one practical reason why 200 nm is the lower wavelength limit of analytical ultra-violet spectrometry (an important application of the vacuum ultra-violet region is considered in § 7.7).

A molecule with two double bonds far apart behaves just like a molecule with one double bond. However, if the double bonds are conjugated, the higher energy levels are lowered and the spectral transition is shifted towards larger wavelengths. As shown

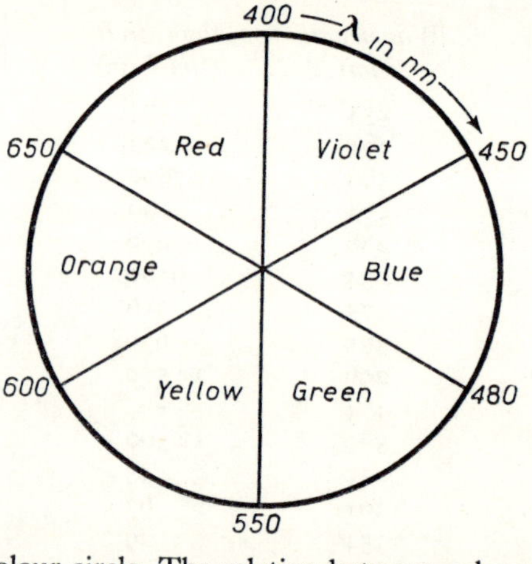

FIG. 4.4. Colour circle. The relation between colour and wavelength (given in nm) is roughly indicated. If radiation of a certain wavelength strikes the eye, the colour impression can be read from the corresponding segment. If a solution absorbs over a wavelength range, the colour of the solution in daylight can be read from the opposite segment (the complementary colour). Naturally, there is a gradual transition from one colour to the next.

in Fig. 4.6, benzene absorbs at 250 nm, and still larger aromatic molecules absorb at increasingly larger wavelengths, as is clear from Table 4.1 (p. 71). Substitutions in the aromatic ring may also shift the absorption band to larger wavelengths.

When the absorption band crosses the border line of 400 nm, part of the visible radiation is absorbed and the compound becomes coloured. This colour may be read from the colour circle in Fig. 4.4. A solution of permanganate absorbs between 500 and 550 nm, that is, green light, and accordingly the diluted solution is pink.

Chelates, important in inorganic analysis (for example EDTA, ethylene diamine tetra-acetic acid), are actually large organic molecules with carboxyl groups, amino groups or aromatic rings. The absorption bands of these molecules mostly lie in the ultra-violet; their stable complexes with many metal ions often show a characteristic colour quite suitable for quantitative spectrometric determinations.

In general, the characteristic line spectrum of a free atom (§§ 2.2 and 7.4) is completely lost in a solution of metal ions. (The spectra of the rare earths, based upon transitions of well-shielded 4*f*-electrons, form a striking exception.) In aqueous solution the cations of copper, iron, chromium etc. form complexes with the surrounding water molecules. In such a large 'molecule' many positions are available to the electron, and broad absorption bands result.

4.5. The Beer–Lambert law

4.5.1. *Derivation*

In Fig. 4.5 a beam of monochromatic radiation of intensity I_0 strikes an absorption cell with an optical path length b, containing a solution with a concentration c of absorbing molecules. After absorption over a distance x in the cell the intensity of the beam is diminished to I. The decrease of the intensity in the adjacent layer dx is proportional to the intensity I and to the number of absorbing molecules (or moles) $c\,dx$ in the layer:

$$-dI = kIc\,dx \tag{4.1}$$

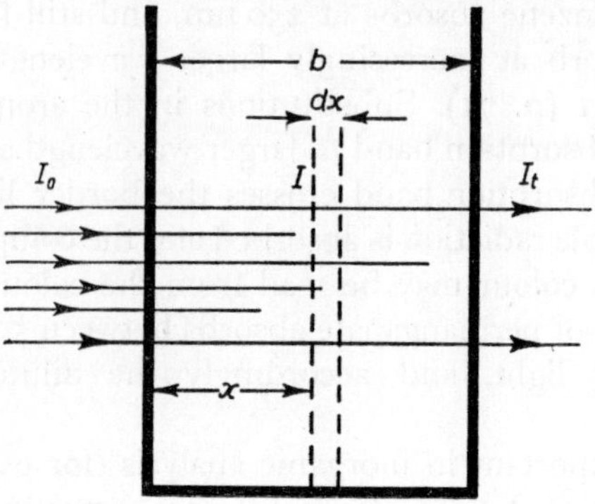

FIG. 4.5. Absorption of radiation in a cell. I_0 = incident intensity, I_t = transmitted intensity. (Further explanation in the text.)

where k is a proportionality constant. Because absorption of radiation results from a spectral transition from the ground state to a higher energy level (§ 1.5), c is strictly speaking the concentration of molecules in the ground state. However, according to Table 1.2 only very few molecules are excited to higher energy levels at room temperature and, thus, c can safely be taken equal to the total concentration. This approximation is also valid for infra-red spectrometry.

With the boundary condition $I = I_0$ for $x = 0$, the solution of the differential equation 4.1 is found to be

$$I = I_0 e^{-kcx}$$

When the absorption cell has been traversed completely, the intensity is reduced to I_t, and the *transmittance T* of the solution is given by

$$T = \frac{I_t}{I_0} = \frac{I(x = b)}{I_0} = e^{-kcb} = 10^{-0 \cdot 4343\, kbc} = 10^{-abc} \quad (4.2)$$

This expression, named after its discoverers as the *Beer–Lambert law*, is of fundamental importance in quantitative absorption spectrometry. It is also applicable to other fields of spectrometry (§§ 6.4 and 8.2).

The transmittance of a solution decreases exponentially with increasing concentration and increasing optical path length *b* of the cell. In addition, the transmittance is determined by the value of the proportionality constant *a*, which is related to the transition probability between two energy levels (§ 2.6) and indicates the basic strength of the absorption band. In older literature this constant is designated by a variety of names, the most popular being *absorption coefficient* (compare §§ 6.4 and 8.2). For molecular absorption spectrometry the term *absorptivity* has been recommended, and this term will be used in the present chapter.

If the concentration is expressed in mol/l and the optical path length in cm, the absorptivity must be expressed in l/mol cm, because the exponent in Equation 4.2 must be of unit dimension. This unit is used in analytical literature. (The SI unit for the absorptivity would be $m^2/kmol$; obviously, $1\ m^2/kmol = 0{\cdot}01$ l/mol cm.) Absorptivities of 10^4–10^5 l/mol cm represent strong absorption bands (large transition probability). Weaker bands have absorptivities of the order of 100–1000 l/mol cm. These often result from transitions which are not allowed by the selection rules of the (inadequate) theory. Accordingly, these are known as 'forbidden' transitions. Obviously, the absorptivity is zero outside the absorption bands. The variation of the absorptivity with wavelength determines the shape of the absorption spectrum.

Because the value of the absorptivity is different at each wavelength and for each particular substance, the Beer–Lambert law is strictly valid for monochromatic radiation only, and applies to each component in the solution separately.

4.5.2. *The absorbance*

The detector of the spectrometer records the intensity of the radiation transmitted by the absorption cell and primarily measures the transmittance, for example as outlined in § 4.1. The inverse, exponential relationship between the transmittance and the concentration is inconvenient in quantitative analysis. Therefore, another quantity is introduced, the *absorbance* A of the solution:

$$A = -\log T = abc \qquad (4{\cdot}3)$$

Numerous experiments have confirmed that the absorbance,

defined in this way, is indeed proportional to the concentration of the absorbing substance. This is expressed by saying that the substance obeys the Beer–Lambert law. However, it is often observed in practice that at larger concentrations the measured absorbance increases more slowly than is expected from Equation 4.3. This is known as a deviation from the Beer–Lambert law.

Strictly speaking, true deviations from this law can only occur if the absorptivity varies with concentration. In molecular absorption spectrometry this only happens at extremely large concentrations, for instance because the refractive index of the concentrated solution changes. Virtually all deviations observed in practice are due to a failure to meet the conditions for which the Beer–Lambert law has been derived. Accordingly, and more appropriately, such deviations are called *apparent* deviations from the Beer–Lambert law. The three most important causes of apparent deviations will be discussed next.

4.5.3. *The influence of the spectral bandwidth*

The Beer–Lambert law has been derived for monochromatic radiation. Actually, however, the monochromator of the spectrometer isolates from the continuum radiation of the source a small wavelength range given by the spectral bandwidth, $\Delta\lambda_s$ (§ 3.2). The intensity of radiation striking the detector in the absence of an absorbing substance is then given by

$$I_0 = \int_{\Delta\lambda_s} I_{0,\lambda}\,\mathrm{d}\lambda = I_{\lambda 0},\ \Delta\lambda_s$$

where $I_{0,\lambda}$ is the source intensity per *unit of wavelength* (compare the definition of the spectral radiance in § 2.7). Over the small range covered by the spectral bandwidth $I_{0,\lambda}$ is practically independent of the wavelength.

Now, when the absorption cell contains an absorbing substance, the intensity striking the detector is given by

$$I_t = \int_{\Delta\lambda_s} I_{0,\lambda}.10^{-a(\lambda)bc}\,\mathrm{d}\lambda = I_{0,\lambda} \int_{\Delta\lambda_s} 10^{-a(\lambda)bc}\,\mathrm{d}\lambda$$

Consequently, the experimentally observed absorbance is

$$A_{\text{exp}} = -\log (I_t/I_0) = -\log \frac{\int 10^{-a(\lambda)bc}\, d\lambda}{\Delta\lambda_s} \tag{4.4}$$

If the absorptivity does not change over the spectral bandwidth, the exponential term can be brought before the integral sign and the experimentally observed absorbance is equal to the absorbance predicted by Equation 4.3.

The spectral bandwidth of ultra-violet spectrometers is typically of the order of 1 nm, whereas the molecular absorption bands are generally smooth and much broader. In most cases, therefore, the influence of the spectral bandwidth will be negligible, especially when the absorbance is measured at the maximum of the absorption band.

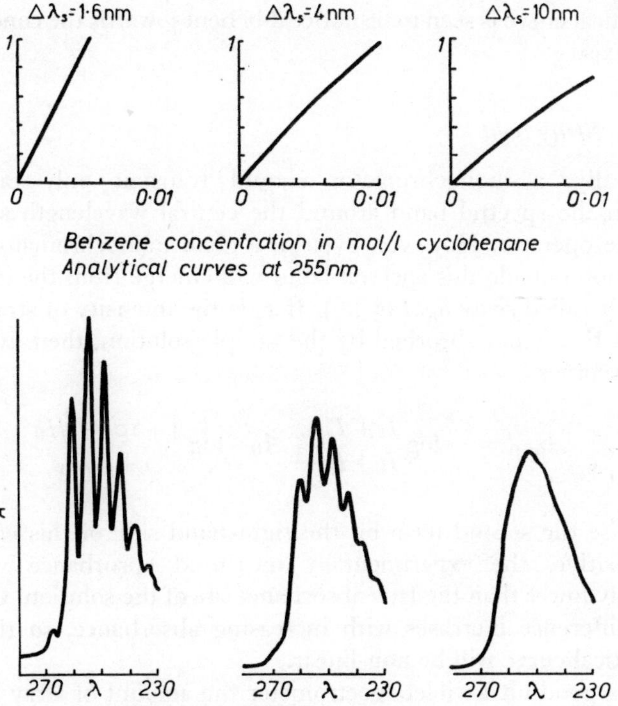

FIG. 4.6. Absorption spectra and analytical curves for benzene at increasing spectral bandwidth.

However, if the absorption band happens to be sharp or measurements are made at the steep slope of an absorption band, the absorptivity may well vary over the spectral bandwidth. In that case, a straightforward solution of Equation 4.4 is impossible, but it is clear that the result also depends upon the concentration c.

As an example, Fig. 4.6 shows the absorption spectrum of benzene, recorded with increasing slit width and hence with increasing spectral bandwidth. Corresponding to each spectrum a graph illustrates the observed relation between the absorbance measured at a fixed wavelength and the concentration of benzene (this is called an *analytical curve*). With increasing slit width the resolving power of the spectrometer decreases and the *recorded* vibration bands gradually blend together. Simultaneously, the absorbance of the maxima decreases and the slope of the analytical curve is reduced. Ultimately, at fairly large spectral bandwidth the analytical curve is seen to be noticeably bent towards the concentration axis.

4.5.4. *Stray light*

Ideally, a monochromator should transmit only radiation within the spectral band around the central wavelength selected by the operator. However, even in a good monochromator some radiation outside this spectral band will emerge from the exit slit. This is called *stray light* (§ 3.6). If I_s is the intensity of stray light and if this is not absorbed by the sample solution, then Equation 4.3 becomes

$$A_{\text{exp}} = -\log \frac{I_t + I_s}{I_0 + I_s} = A_0 - \log \frac{1 + 10^{A_0} I_s/I_0}{1 + I_s/I_0} \qquad (4.5)$$

Because the second term on the right-hand side of this equation is positive, the experimentally measured absorbance will be slightly lower than the true absorbance A_0 of the solution. Clearly, the difference increases with increasing absorbance, so that the analytical curve will be non-linear.

In a good ultra-violet spectrometer the amount of stray light is generally very small ($I_s/I_0 < 10^{-3}$), so that this effect is negligible. In the far ultra-violet region and in the infra-red (§ 5.1) stray light can become appreciable and lead to noticeable bending of the analytical curves.

4.5.5. *Influence of chemical equilibria*

In Equation 4.3, c represents the true concentration of the absorbing substance. Practical analysis often deals with substances that are in equilibrium with other components of the solution. Any shift of this equilibrium with increasing concentration will produce a change of the concentration c, and a corresponding change of the observed absorbance. A simple example is offered by a compound IH, that is partly dissociated into I and H. If c is the total concentration, α is the degree of dissociation, and K is the dissociation constant, the following expressions can be written down:

$$[\mathrm{HI}] = (1-\alpha)c$$
$$[\mathrm{H}] = [\mathrm{I}] = \alpha c$$
$$K = \alpha^2 c/(1-\alpha)$$

And for nearly complete dissociation ($\alpha \simeq 1$),

$$\alpha = 1 - c/K$$

If a wavelength is chosen, where I absorbs, but IH does not, then the absorbance is

$$A_{\mathrm{exp}} = ab[\mathrm{I}] = ab\alpha c = ab(1-c/K)c = A_{\mathrm{th}}(1-c/K)$$

With increasing concentration the experimentally observed absorbance increasingly lags behind the absorbance expected theoretically from Equation 4.3, and the analytical curve is bent towards the concentration axis. Chemical interferences of this kind occur rather frequently. The interference discussed here is effectively removed by the addition of excess H (buffered solution), which makes the concentration of free I a constant fraction of the analytical concentration c (see problem 7).

4.6. Quantitative analysis

4.6.1. *Analytical curves*

Equation 4.3 seems to indicate that the solution concentration of a substance can be calculated directly from the measured absorbance, once the absorptivity of the substance is known. It appears, however, that the absorptivity is sensitive to such factors as temperature, pH, solvent and even the spectrometer used in the determination. Therefore, values for the absorptivity quoted in the literature (for example in Table 4.1) cannot be used for accurate quantitative analysis. Rather, a relative method is followed and the concentration of the unknown sample solution is read from an analytical curve periodically determined with a series of standard solutions of known concentration.

Generally, a blank determination is carried out with a solution closely resembling the sample solution, but without the constituent to be determined. Obviously, standards, sample and blank solutions must be analysed under the same conditions and treated in the same way (heating, extraction, reagent concentration, buffer, etc.). In order to correct for radiation losses through reflection or absorption in the cell windows all solutions are measured relative to a reference solution, for example the pure solvent.

If the substance possesses a strong absorption band in the ultraviolet or visible region this may be used directly for quantitative analysis (aromatics, permanganate, dichromate). In other cases a reagent is added that forms an absorbing compound with the constituent in question. Large organic molecules often serve as *colour reagents* in inorganic analysis. Occasionally the colour reaction proceeds instantaneously, but usually it takes several minutes for the colour to develop completely and sometimes heating is necessary. For any determination it is essential that the absorbance remains stable for at least half an hour to allow sufficient time for the actual absorption measurement. Also, the absorbance should be reproducible and insensitive to minor variations of pH, temperature, amount of reagent added, etc.

Preferably, the absorbing compound should obey the Beer–Lambert law, because a linear analytical curve is conveniently determined and checked. However, this is by no means essential and slightly bent analytical curves can be perfectly suitable for quantitative analysis. In fact, when inspected closely the majority of analytical curves are non-linear.

Absorptivities of the order of 1000 l/mol cm are generally sufficient for analytical determinations, but for the measurement of small concentrations (trace analysis) the absorptivity should obviously be as high as possible (up to 100000 l/mol cm). For the determination of very small amounts (microanalysis) specially designed cells are used that combine a small cell volume (0·2 ml) with a long optical path (5 cm).

The major problem in molecular absorption spectrometry is the occurrence of interferences from other constituents of the sample. It is obviously desirable that the determination be free from interferences so that sample pretreatment is reduced to a minimum. In practice, however, interferences are the rule rather than an exception. Other constituents of the sample may absorb at the wavelength of interest. Another ion may form an absorbing compound with the reagent (no reagent is *specific*). The component in question may be strongly bonded with anions that prevent its reaction with the reagent. As a result, developing a new spectrometric absorption determination is time-consuming and may involve a lot of 'chemistry'. Separation procedures, notably solvent extraction, often precedes the measurement of the absorbance.

However, once the analytical procedure has been established, ultra-violet and visible absorption spectrometry offers the advantage of a *simple*, *precise* and *fairly rapid* determination that can be carried out with moderately priced instruments. In marked contrast to many other spectrometric determinations a molecular absorption analysis can often be adapted or taken directly from the literature without much trouble. In fact, it is usually possible to apply one of the numerous determinations that have been described in the literature or collected in the handbooks. By the introduction of new colour reagents the number of absorption procedures is continuously expanding, and it would be hard to find a field where this form of spectrometric analysis is not being utilized.

4.6.2. *Precision and limit of detection*

An expression for the relative error in the determination of the concentration can be derived from Equation 4.3:

$$\frac{\mathrm{d}c}{c} = \frac{\mathrm{d}A}{A} = \frac{\mathrm{d}\log T}{\log T} = \frac{0{\cdot}43\,\mathrm{d}T}{T\log T} \tag{4.6}$$

Apparently, the relative error in the concentration is determined by the error in measuring the transmittance (dT) and by the magnitude of the transmittance. Fig. 4.7 shows the relative error as a function of the transmittance for two cases. In curve (a) it is assumed that the error in the transmittance is equal over the entire scale of the detection instrument: $dT = 0{\cdot}0035$. In curve (b) it is assumed that the detector produces a random error proportional to the square root of the incident intensity. It then follows from

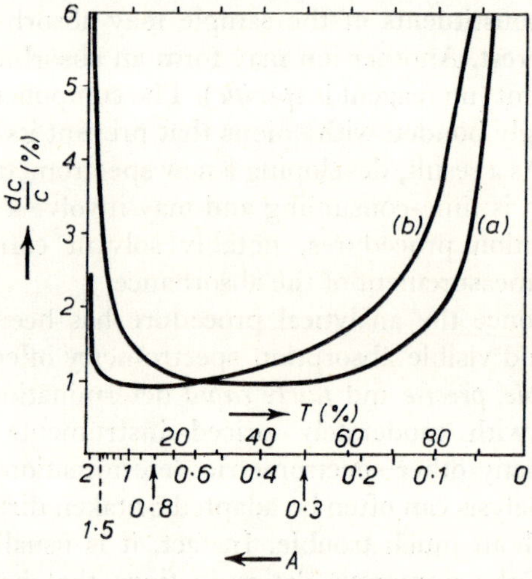

FIG. 4.7. Relative error in the measurement of absorbance. Curve (a), constant error in the transmittance; curve (b), error proportional to the square root of the transmittance. In drawing these curves it is assumed that in both cases the relative error is equal to 1 per cent for an absorbance of 0·6.

Equation 4.2 that the error in the transmittance, dT, is proportional to $\sqrt{T}$. For both curves it is *arbitrarily* assumed that the relative error in the concentration is equal to 1 per cent for a transmittance of 25 per cent ($A = 0{\cdot}6$).

In both cases the relative error in the concentration is seen to change little in the absorbance range between 0·1 and 1·5, but to increase rapidly for higher and lower absorbances. In practice, the error curves of most spectrometers lie somewhere between these theoretical curves. Preferably, solutions are prepared in such a way that the absorbance falls within the range 0·1–1·5 (or the even more restricted range of 0·3–0·9). In this optimum range the precision of a molecular absorption determination is found to be about 0·5 per cent, although better results are sometimes reached.

As is readily seen from Equation 4.3, a compound with a strong absorption band ($a = 10^5$ l/mol cm) may be determined in a 1 cm cell over a concentration range of 10^{-5}–10^{-6} mol/l. For weaker absorption bands the concentration must be raised appropriately.

The limit of detection (the minimum concentration that can be significantly detected) not only depends upon the value of the absorptivity, but also upon the noise level of the detection instrument. With most spectrometers an absorption of 1 per cent ($T = 0{\cdot}99$, $A = 0{\cdot}004$) can still be measured. Hence, an absorptivity of 10^5 l/mol cm and a 4 cm absorption cell allow the detection of concentrations down to 10^{-8} mol/l.

4.6.3. *Multicomponent analysis*

Up to now it has been assumed that only one absorbing component is present in the sample solution. Especially in a mixture of organic compounds it will often be impossible to select a wavelength where only one substance absorbs. Instead of resorting to tedious and time-consuming separations, it is sometimes possible to determine the components simultaneously. Because absorbances are *additive*, the absorbance of a mixture of absorbing substances is equal to the sum of the separate absorbances.

For a mixture of two components P and Q, let the absorbance be measured at two wavelengths, λ_1 and λ_2.

$$\begin{aligned} A(\lambda_1) &= a_{P,\lambda_1} bc_P + a_{Q,\lambda_1} bc_Q \\ A(\lambda_2) &= a_{P,\lambda_2} bc_P + a_{Q,\lambda_2} bc_Q \end{aligned} \qquad (4{\cdot}7)$$

where a_{P,λ_1} is the absorptivity of the pure substance P at wavelength λ_1. Together with the other three absorptivities this quantity is derived from spectra of the pure components P and Q. The two simultaneous equations 4.7 can then be solved for the unknown concentrations c_P and c_Q, once the two absorbances have been measured.

This procedure gives satisfactory results for two or three components, but gets increasingly complicated for more complex mixtures. Obviously, the number of equations must at least be equal to the number of sample constituents, and for numbers greater than five an electronic calculator must be used to perform the calculations. Also, the wavelengths must be chosen with care. Preferably, each component should in turn give a major contribution to the absorbance.

In general, this is not feasible for complicated mixtures and, consequently, the precision of multicomponent analyses is markedly less than the precision attained in single-component analyses.

4.7. Photometric titrations

Colour indicators are frequently used to show the endpoint of a titration. In some complexometric titrations it is difficult to observe the colour change of metal indicators visually. It can then be advantageous to use instrumental detection. If the indicator changes from red to blue, the spectrometer may be set to a wavelength of 650 nm. Under favourable conditions, the absorbance of the titrated solution then shows a sudden increase from an extremely low value to a rather high value at the endpoint of the titration. Instrumental detection also offers the possibility of automation. For instance, the titrant can be added mechanically and the supply can be stopped automatically when a preset absorbance has been reached. The concentration of the titrated constituent may then be calculated (electronically) from the amount of solution added.

For so-called *self-indicating* titrations instrumental detection offers completely different possibilities. In the titration of ferrous ions with permanganate the endpoint is observed visually by the first appearance of a pink colour. If for this reaction the absorbance of the solution is measured at 525 nm, the absorbance will remain zero until the equivalence point. After this point the absorbance increases linearly with the excess permanganate added. Therefore, a plot of the absorbance as a function of the amount of permanganate solution added yields two straight lines, and the point of intersection corresponds to the endpoint of the titration (Fig. 4.8).

Such *photometric* titrations offer several advantages over conventional, visual titrations. Because the straight lines are based upon many points, the equivalence point can be determined very precisely. Since the titration is continued after the equivalence point, no droplet error occurs. When low concentrations are to be titrated, the reaction may be incomplete near the equivalence point, but the true equivalence point can still be found by a simple extrapolation of the linear parts of the titration curve (see Fig. 4.8).

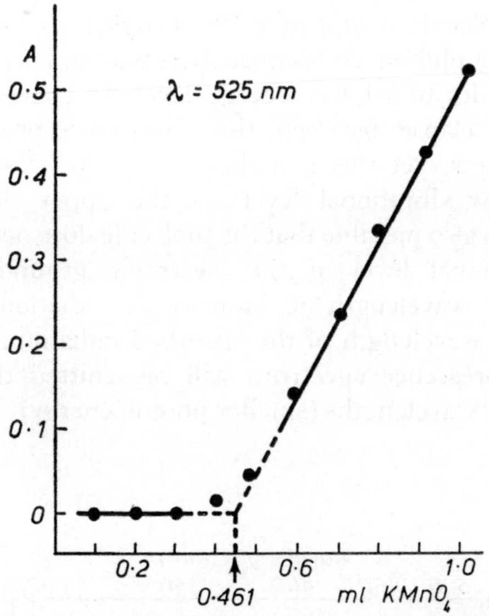

FIG. 4.8. Photometric titration of ferrous ions with permanganate.

Concentration of permanganate solution: 0·003 M
equivalence point: 0·461 ml
total cell volume: about 10 ml
amount of iron present: 0·386 mg
absorptivity of permanganate: 2000 l/mol cm

Photometric titrations are often preferred over analytical curve procedures, because the titration is not hindered by the presence of other absorbing compounds and no time is lost in preparing the analytical curve (a procedure that must be repeated from time to time). On the other hand, a photometric titration is only possible if the titration reaction proceeds rapidly. Further, the absorbance must change at the equivalence point and the absorbing substances must obey the Beer–Lambert law.

4.8. Fluorimetry

4.8.1. *Instrumentation*

In § 1.5 *fluorescence* was introduced as a possible radiation process. A molecule is first raised to a higher energy level by the absorption of a photon. Subsequently, a photon is emitted and the molecule recedes to a lower energy level. In general, only a few microseconds elapse between the absorption process and the emission process, yet this is sufficient to allow the molecule to descend a few vibrational levels in the upper electronic level (Fig. 2.3). It is also possible that the molecule does not return to the lowest vibrational level of the electronic ground state. Consequently, the wavelength of fluorescence radiation need not be similar to the wavelength of the absorbed radiation. In general, a complete fluorescence spectrum will be emitted that is shifted towards larger wavelengths (smaller photon energy). This is shown in Fig. 4.9.

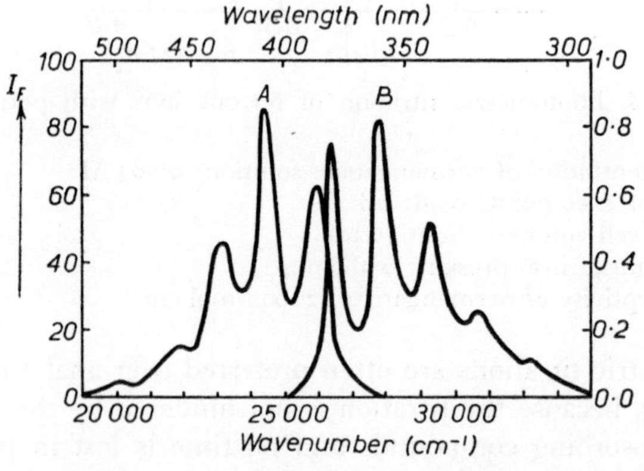

FIG. 4.9. Fluorescence spectrum (A) and absorption spectrum (B) of anthracene.

An instrument for the excitation and the measurement of fluorescence radiation is shown schematically in Fig. 4.10. *Excitation radiation* of a particular wavelength is selected by monochromator M_1 from the spectrum of the source S and directed to the sample solution in a cell with four optical windows. In order to prevent excitation radiation from reaching the detector, the second monochromator M_2 is placed at a 90° angle. Fluorescence radiation is emitted in all directions, and part of this enters the monochromator M_2, where it is dispersed and detected in the usual way.

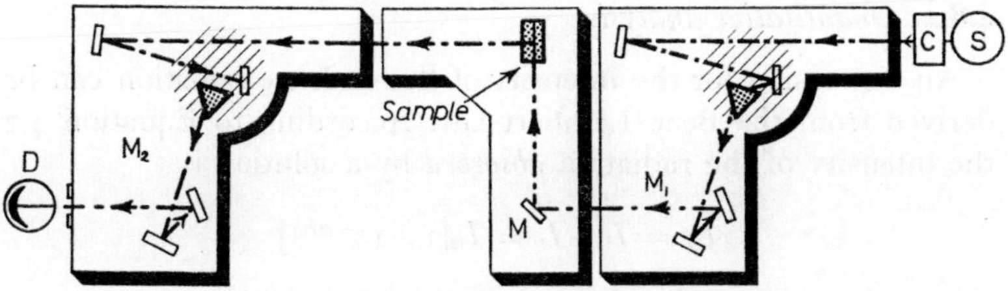

FIG. 4.10. Optical arrangement of a spectrofluorimeter (by courtesy of Hilger and Watts).

S = source of radiation
C = lens
M_1 = monochromator for isolating the excitation radiation
M = mirror
M_2 = monochromator for analysing the fluorescence spectrum
D = detector

The measurement of fluorescence radiation proceeds rather inefficiently. The absorbed radiation is not completely converted to fluorescence radiation and only a small fraction of this radiation enters the monochromator M_2, where it is distributed over several wavelengths. Therefore, the source of radiation should be very intense; a xenon arc lamp is frequently applied (§ 4.3). With this continuum source of radiation and an arrangement of two monochromators the wavelength of both the excitation radiation and the fluorescence radiation can be varied at will. This is important for fundamental investigations, such as a study of the position of vibrational levels in the electronic ground state and higher energy levels of molecules.

Because relatively few substances exhibit fluorescence activity, the structure of an observed fluorescence spectrum often provides information about the nature of the compound studied. In that case a simpler instrument may suffice, where the monochromator M_1 is replaced by a filter, and a mercury discharge lamp is used as the source of radiation. One of the intense mercury lines isolated by the filter is then used as the excitation radiation.

In even simpler instruments the second monochromator is also replaced by a filter. Such instruments can only be utilized for quantitative analysis.

4.8.2. *Quantitative analysis*

An expression for the intensity of fluorescence radiation can be derived from the Beer–Lambert law. According to Equation 4.2 the intensity of the radiation *absorbed* by a solution is

$$I_a = I_0 - I_t = I_0[1 - 10^{-abc}]$$

As discussed in § 1.5 the energy gained by the absorption of radiation is usually converted into heat through molecular collisions. However, if a substance shows fluorescence activity, part of the absorbed energy is re-emitted as radiation. Consequently, fluorescent emission and *quenching* collisions are competitive processes and the fluorescence intensity is only a fraction Φ of the absorbed intensity:

$$I_F = \Phi I_a = \Phi I_0[1 - 10^{-abc}] \qquad (4.8)$$

Φ is called the *fluorescence yield.* In favourable cases it can reach values between 0·1 and 0·6. (Of course, most substances show no fluorescence activity, which means that Φ is zero.)

For very small concentrations Equation 4.8 can be written as

$$\begin{aligned} I_F &= \Phi I_0[1 - e^{-2\cdot 3abc}] \\ &= \Phi I_0[1 - \{1 - 2\cdot 3\, abc + (2\cdot 3\, abc)^2/2 + \ldots\}] \\ &= 2\cdot 3\, \Phi I_0\, abc[1 - 1\cdot 15\, abc + \ldots] \end{aligned}$$

If the absorptivity is equal to 10^4 l/mol cm and a 1 cm cell is used, the fluorescence intensity of concentrations smaller than 10^{-6} l/mol may be approximated as

$$I_F = 2{\cdot}3\,\Phi I_0\,abc \tag{4.9}$$

Consequently, the fluorescence intensity is only proportional to the concentration so long as the product *abc* is smaller than 0·1 (see problem 12). Indeed, quantitative fluorimetric analysis is normally applied to the determination of low concentrations. This would tend to make the fluorescence intensity very weak, but Equations 4.8 and 4.9 show that the fluorescence intensity is also proportional to the excitation intensity I_0 of the source. For this reason high intensity mercury discharge lamps or xenon arc lamps (§ 4.3) are applied to increase the sensitivity of the determination. Limits of detection down to 10^{-10} mol/l may then be reached, and fluorimetric analysis is frequently used for trace analysis.

The precision of the analysis is less than for absorption spectrometry and amounts to 1–3 per cent. In addition, fluorimetry is rather sensitive to interferences. Minor concentrations of a strongly fluorescent compound may obliterate the signal of a weakly fluorescent compound. Quenching collisions may be promoted by the presence of other components (dissolved oxygen gas is a notorious example) and seriously reduce the fluorescence yield Φ. Preliminary solvent extraction is frequently required and the analysis must be carried out with great care.

Quantitative fluorimetry is mainly applied to the analysis of biological materials (tissues, pesticides, vitamins, enzymes, etc.). Inorganic trace analysis is feasible with fluorescence reagents that form fluorescent compounds with metal ions.

Bibliography

1. BAUMAN, R. P., *Absorption Spectroscopy*, Wiley, New York, 1962.
2. EDISBURY, J. R., *Practical Hints on Absorption Spectrometry*, Adam Hilger, London, 1966.
3. HEADRIDGE, J. B., *Photometric Titrations*, Pergamon Press, Oxford, 1961.
4. HERCULES, D. M., ed., *Fluorescence and Phosphorescence Analysis*, Interscience, New York, 1966.

5. DONBROW, M., *Instrumental Methods in Analytical Chemistry; Part II, Optical Methods*, Pitman and Sons, London, 1967.
6. CALDER, A. B., *Photometric Methods of Analysis*, Adam Hilger, London, 1969.
7. KORTÜM, G., *Kolorimetric, Photometric und Spektrometric*, 4th ed., Springer, Berlin, 1962.

Problems

1. In a particular absorption determination pure solvent is used as the reference. Two solvents are available that are equally suitable from a chemical point of view. With respect to air their transmittances are 70 per cent and 10 per cent respectively. Which solvent should be preferred, and why?

2. The intensity of a hydrogen discharge lamp is approximately constant between 300 and 200 nm. A double-beam spectrometer with a grating monochromator operates according to the principle of constant-energy detection. If it is observed that the slit width increases steadily towards smaller wavelengths, what conclusion can be drawn about the detector?

3. Can an acetate buffer be used, if absorptions are to be measured around 230 nm?

4. What conclusion would you draw from the presence of a very weak absorption band at 250 nm in the spectrum of cyclohexane?

5. The absorptivity of caffeine ($C_8H_{10}N_4O_2$) at 278 nm is 9200 l/mol cm. How many grams must be dissolved in 100 ml water to obtain a transmittance of 50 per cent in a 1 cm absorption cell? What is the transmittance of this solution in a 4 cm cell?

6. At 450 nm the following absorbances are measured for solutions of riboflavin ($M = 376$):

Concentration mg/l	*Absorbance in 1 cm cell*
5	0·142
10	0·284
20	0·565
30	0·840
40	1·105
50	1·37

Draw the analytical curve and calculate the absorptivity. If the precision of the spectrometer agrees exactly with curve *a* in Fig. 4.7, can it be concluded that riboflavin obeys the Beer–Lambert law?

7. Let K denote the dissociation constant of an indicator HI. The absorbance of the acid form (I) is measured in buffered solutions of equal pH. Show that the absorbance can be expressed as

$$A = \frac{abc}{1+[H]K}$$

where c is the total concentration of indicator added.

8. Give a qualitative explanation of Fig. 4.6.

9. The spectra of three solutions consist of single, broad bands with maxima at 500, 600, and 700 nm, respectively. What is the colour of these solutions? If these substances are to be determined quantitatively with a filter photometer, what colour will be chosen for the filter?

10. By differentiating Equation 4.6 find that value of the transmittance for which the relative error in the concentration is least. Recall that in curve a of Fig. 4.7 dT is constant, whereas in curve b it is assumed that $dT = p\sqrt{T}$, where p is constant.

11. The photometric titration of ferrous ions with permanganate in Fig. 4.8 is an example of the general case $A+B \rightarrow C$, where A and C are colourless and B absorbs. Draw graphs for the following cases:

(*a*) A absorbs, B and C colourless
(*b*) A and B absorb, C colourless
(*c*) B and C absorb, A colourless

What condition must be fulfilled in case (*c*) to allow photometric titration?

12. Equation 4.9 represents the analytical curve for fluorimetric analysis of small concentrations. Draw the complete analytical curve represented by Equation 4.8. For example, choose $a = 10^4$ l/mol cm, $b = 1$ cm and c varying from 10^{-8} to 10^{-2} mol/l. Use double-logarithmic graph paper and check the linear portion of the curve.

13. A series of solutions have true absorbances ranging from zero to 1·5. Use Equation 4.5 to calculate the absorbances observed experimentally in the presence of 1 per cent stray light ($I_s/I_0 = 0{\cdot}01$). Plot the results on a graph to note the curvature of the analytical curve.

CHAPTER FIVE

INFRA-RED SPECTROMETRY

5.1. Radiation sources and detectors for infra-red spectrometry

The sources of radiation discussed in § 4.3 are not suitable for infra-red spectrometry. The glass or silica envelopes required to contain the low-pressure or high-pressure gas do not transmit infra-red radiation. Moreover, it is clear from Fig. 2.5 that sources with a high radiation temperature ($T > 2500$°K) predominantly emit radiation of smaller wavelength. Even a small fraction of short-wavelength radiation would generate a much greater signal in the infra-red detector than the infra-red radiation proper. Even in a good monochromator this *stray light*, due to reflections from mirrors, gratings and polished metal slits, cannot be entirely prevented.

Therefore, radiation sources for infra-red spectrometry simply consist of a glowing rod, placed free in air and heated electrically to a temperature of 1500–2000°K. The *Nernst* source is made from a mixture of zirconium oxide (90 per cent), yttrium oxide (7 per cent) and erbium oxide (3 per cent). At temperatures around 2000°K the rod is electrically conductive. The lifetime of this source varies from several months to one year. The *globar* consists of a carborundum (SiC) rod, which is conductive at room temperature and is heated to a temperature of 1500°K.

According to Fig. 2.5 the intensity of these continuum sources is rather small, and decreases rapidly with increasing wavelength. Therefore, the detector must be extremely sensitive. The energy of infra-red photons is too small to yield photoelectrons. Photoconductive cells operated at room temperature (§ 4.3) can only be used for wavelengths up to 5 μm. Photoconductive cells for longer

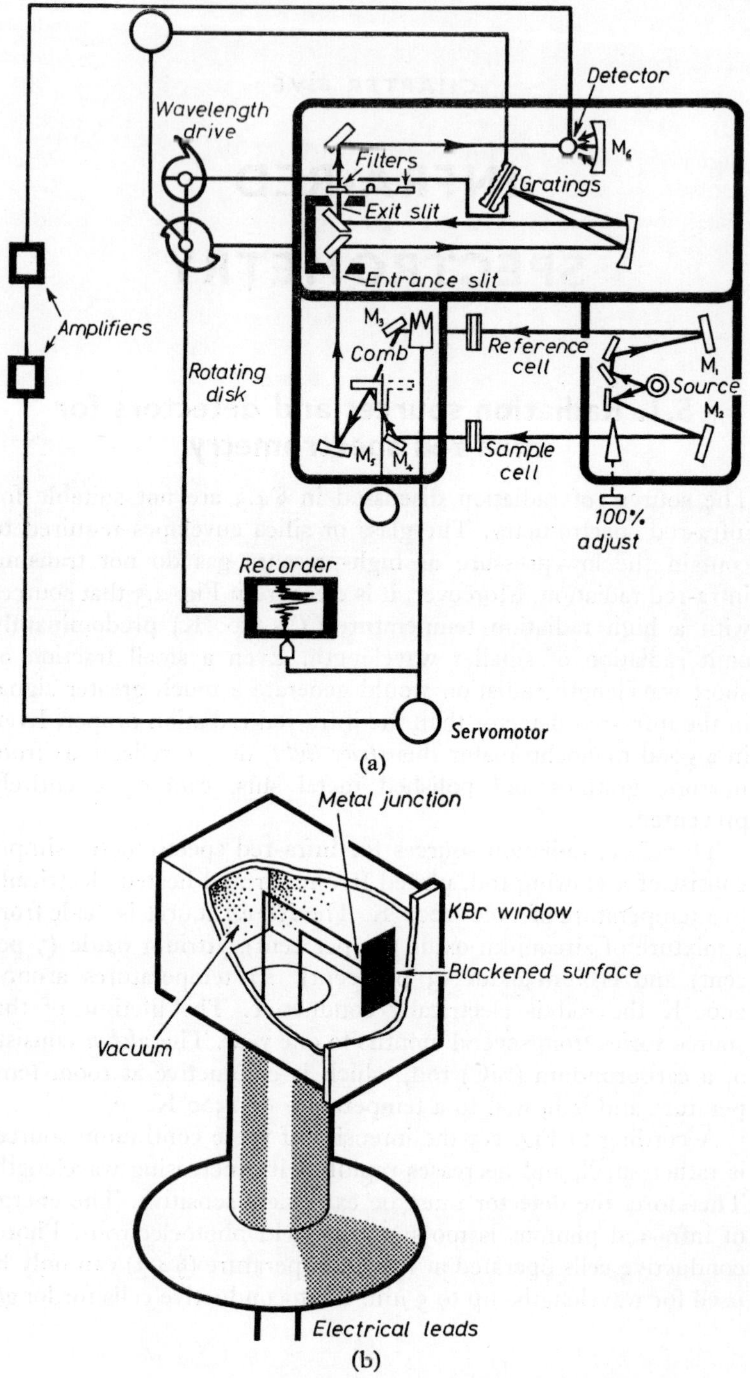

FIG. 5.1. (a) Schematic diagram of an infra-red spectrometer (model EPI–L from Hitachi, Japan). (b) Construction of a thermocouple.

wavelengths must be cooled to at least 200°C below freezing point and are therefore not used in analytical instruments.

By far the most popular infra-red detector is the *thermocouple* (Fig. 5.1b) which consists basically of a junction of two dissimilar metals (e.g. gold and tellurium). Incident radiation is effectively absorbed by a small blackened gold leaf attached to the junction, and the energy of the radiation is converted to heat. The rise in temperature causes a change in the potential difference across the junction, which can be readily measured. In the *thermistor bolometer* the temperature increase from absorbed radiation changes the resistance of a thin film of metal oxides, and this can be measured by a Wheatstone bridge circuit.

Now, the incident infra-red radiation produces a temperature increase of only 0·01°C, and provision must be made to distinguish the effect of the radiation from the much larger variations due to the fluctuating room temperature. For this reason the incident radiation is *chopped* with a frequency of about 10 Hz. The interrupted radiation produces a small alternating current, which is selectively amplified with a tuned alternating current preamplifier placed directly behind the infra-red detector. The much larger, but also slower voltage changes due to the fluctuating room temperature are rejected by the preamplifier, and only the 10 Hz signal is transmitted.

5.2. Infra-red spectrometers

The infra-red absorption spectrum is an important tool for the identification and structural analysis of an organic compound. For this application fully automatic double-beam spectrometers have been developed; an example is shown in Fig. 5.1a. It is instructive to compare this diagram with the one in Fig. 4.2.

The concave mirrors M_1 and M_2 reflect two beams of radiation from the source. One beam traverses the reference cell, is deflected by the plane mirror M_3 and passes through the open portion of the rotating disk onto the concave mirror M_5. The other beam traverses the sample cell and is deflected by the plane mirror M_4 and the mirror portion of the rotating disk towards M_5. The two beams alternately enter the monochromator where they are dispersed by the grating.

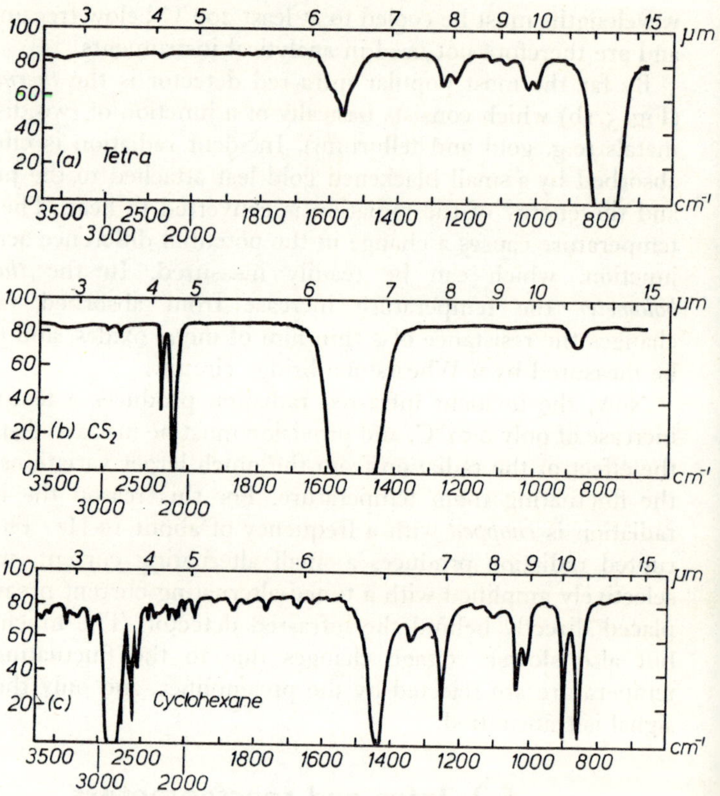

FIG. 5.2. Infra-red spectra

An elliptical mirror M_6 is used to focus the beam of radiation emerging from the exit slit exactly onto the very small photosensitive surface of the thermocouple (0·6 mm^2). If placed behind the exit slit, absorption cells would disturb the exact focusing of the beam, and, therefore, in an infra-red spectrometer the absorption cells will always be found between the source and the entrance slit of the monochromator. In view of the smaller photon energy and lower source intensity, decomposition of sample molecules need not be feared (compare Fig. 4.2).

Two gratings provide an optimum resolving power over the entire infra-red region. One grating, with 200 grooves per mm (d = 5 μm) and blazed for 3·85 μm, is suitable from 2·5 to 8 μm;

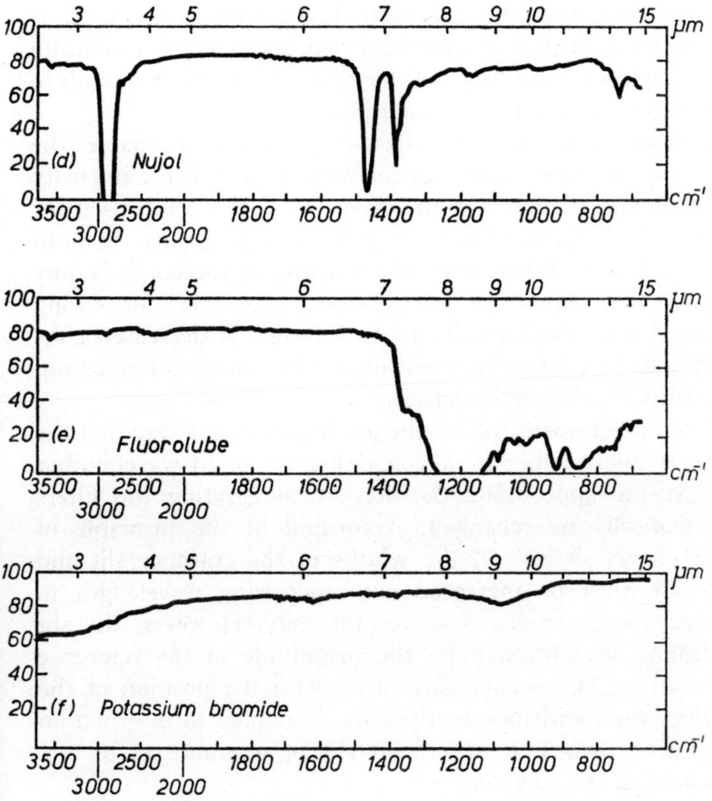

olvents and suspending agents.

the other grating, with 66 grooves per mm ($d = 16\ \mu m$) and blazed for 11·5 μm, is used from 7·5 to 25 μm. Separation of the various orders is achieved by four filters.

In recording the infra-red absorption spectrum the principle of the *optical-null* is applied. When the sample solution absorbs radiation, the two beams of radiation entering the monochromator are unbalanced and the intensity striking the detector varies with a frequency determined by the speed of revolution of the chopper (10 Hz). The resulting alternating voltage from the detector is amplified and used to drive a comb (Plate 2) into the reference beam. As the comb moves into the beam, the reference intensity is attenuated until the intensities of both beams are equal and the force driving the comb becomes zero. The position of the comb is coupled to the deflection of a recorder pen. At the start of each run the beam intensities are balanced by manual control of a comb in the sample beam (100 per cent adjustment).

For measurements according to the optical-null principle, the detector response need not be linear with respect to the intensity of the incident radiation. On the other hand, the comb used for attenuating the reference beam must be of high quality. At both ends of the transmittance scale, the linearity of the comb is only fair, and the readings are less reliable. Moreover, for strong absorption bands the intensity of *both* beams is decreased substantially and the detector receives only a small amount of radiation, which results in a loss of sensitivity.

A driving mechanism rotates the grating in such a way that the spectrum is linear either in wavelength (μm) or in wavenumber (cm^{-1}). At the appropriate position of the grating the filters are automatically interchanged. According to the principle of 'constant energy detection', the widths of the entrance slit and the exit slit must be increased with increasing wavelength to compensate for the reduced source intensity. However, the slit width cannot be controlled by the magnitude of the reference signal (as in § 4.2), because this varies with the position of the comb. Therefore, widening of the slits is coupled to the rotation of the grating according to a fixed slit programme set by the manufacturer of the spectrometer.

5.3. Solvents and absorption cells

Preferably, the sample should be in the solution state. Unfortunately, every molecule with more than two atoms shows vibrational modes that can be excited by suitable infra-red radiation. Consequently, no solvents are available that are completely transparent over the entire infra-red region. The consequences of solvent absorption may be evaluated from Equation 4.3, although, strictly speaking, this expression does not apply to pure liquids. A pure liquid with a density of 1 g/cm^3 and a molecular mass of 100 corresponds to a concentration of 10 mol/l. Even in a thin layer of 0·001 mm a normal vibration band with an absorptivity of 10^3 l/mol cm will then yield an absorbance equal to $A = 1$. In a 0·1 mm absorption cell the transmittance of pure solvents will be smaller than 1 per cent over a broad spectral region. In that case the spectrometer no longer operates properly. The amount of radiation transmitted by the cells is insufficient to activate the detector, the amplifier receives no signal, and the recorder is 'dead'. No significance at all can be attached to the position of the recorder pen.

Several provisions are made to solve this problem. Solvents are selected from compounds consisting of small, symmetrical molecules with only a few strong absorption bands in the infra-red region. As is clear from Figs. 5.2a and 5.2b, carbon tetrachloride shows only one strong absorption band at 700 cm^{-1}, whereas carbon disulphide is completely transparent for wavenumbers below 1300 cm^{-1} ($\lambda > 7{\cdot}5$ μm). This suggests the following technique for obtaining the infra-red spectrum of a substance: a solution in CCl_4 is used for the region between 4000 and 1300 cm^{-1}, and an equally concentrated solution in CS_2 is used for smaller wavenumbers. Obviously, this procedure can only be applied to substances that are sufficiently soluble in both solvents. For polar compounds chloroform is substituted for carbon tetrachloride, whereas for amines, which react chemically with carbon disulphide, cyclohexane is used as a solvent for wavenumbers below 1300 cm^{-1}.

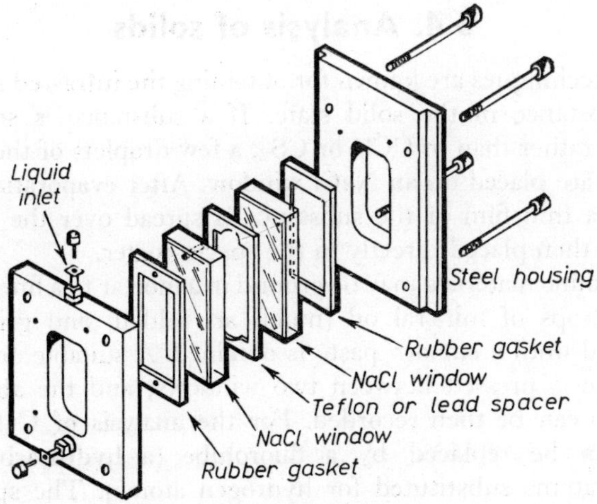

FIG. 5.3. Construction of an infra-red absorption cell.

The influence of weak absorption bands from the solvent is reduced by using absorption cells with a small optical path length. Because glass and silica do not transmit infra-red radiation (Fig. 3.3), the cell windows are made of NaCl (applicable down to 650 cm^{-1}), KBr (down to 400 cm^{-1}) or CsBr (down to 250 cm^{-1}). The construction of an infra-red absorption cell is shown in Fig. 5.3. The optical path length of the cell is determined by the thickness of the spacer of lead or Teflon placed between the cell windows. Protective Teflon gaskets are inserted between the brittle windows and the metal cover plates. All parts are held together by metal bolts.

Solutions are injected between the windows by means of a syringe through holes drilled in the front parts of the cell. The optical path length usually ranges from 0·005 to 0·1 mm; accordingly, rather concentrated solutions are required (1–10 per cent) to obtain a good absorption spectrum. Liquid compounds may be introduced directly into the absorption cell. This avoids the disadvantages of using solvents, namely limited solubility, solvent absorption and physical or chemical interaction between sample and solvent molecules. On the other hand, the optical path length must be reduced to 0·01 mm or less, and the sample molecules may strongly interact mutually, which changes the spectrum.

5.4. Analysis of solids

Several techniques are known for obtaining the infra-red spectrum of a substance in the solid state. If a substance is soluble in acetone, rather than in CCl_4 or CS_2, a few droplets of the acetone solution are placed on an NaCl window. After evaporation of the acetone a thin film of the substance is spread over the window, which is then placed directly in the spectrometer.

Crystalline materials may be ground in a mortar to a fine powder. A few drops of mineral oil (nujol) are added, and grinding is continued until a smooth paste is obtained. A suitable amount of this paste is pressed between two windows, and the absorption spectrum can be then recorded. For the analysis of C–H bands nujol can be replaced by a fluorolube (a hydrocarbon with fluorine atoms substituted for hydrogen atoms). The spectra of nujol and fluorolube are presented in Fig. 5.2.

The finely powdered crystalline substance can also be mixed with potassium bromide, and the mixture formed into a clear tablet by a pressure of a few tons per cm^2. This technique offers the advantage that KBr does not absorb in the infra-red region.

A completely different technique has been developed to obtain the infra-red spectrum of amorphous, insoluble solid materials. The basic principle is shown in Fig. 5.4. Infra-red radiation enters a crystal of high refractive index (for example germanium, $n = 4$). The sample (a paste, a paint, a piece of cotton, etc.), which

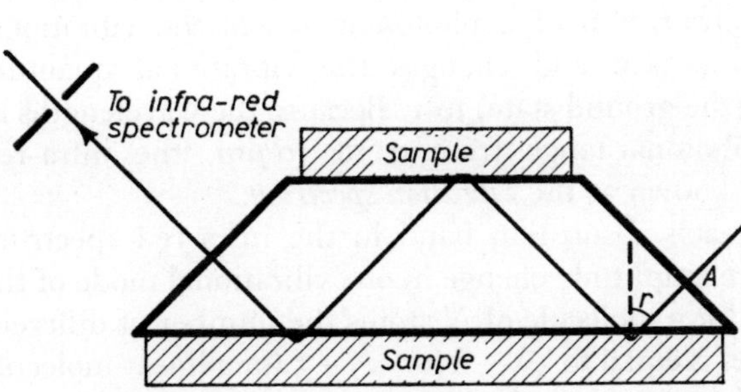

FIG. 5.4. Principle of the ATR technique (see the explanation in the text).

has a much lower index of refraction, is closely attached to both sides of the crystal. For suitable angles of incidence the beam of radiation is *totally reflected* at the boundaries between crystal and sample (§ 1.1). However, in doing so the beam enters the sample for an extremely small distance, and may be absorbed. This is called *attenuated total reflection* (ATR). The effect is greatly enhanced by letting the beam reflect back and forth several times. One obvious advantage of the ATR technique is that only the surface of the sample is analysed. For example, the technique may be applied to the infra-red analysis of polymer coatings on razor blades.

In all cases the spectra of solid samples strongly resemble the spectra of the corresponding solutions. Small shifts and splitting of absorption bands may result from changes in the vibrational modes of the atoms due to the regular arrangement of the molecules in a solid. In view of the limited reproducibility of sample preparation, however, none of the techniques discussed in this section is suitable for accurate quantitative analysis.

5.5. Infra-red absorption and molecular symmetry

As has been discussed in § 2.4, the atoms in a molecule oscillate around an equilibrium position, and the amplitude of this vibration may be altered abruptly under the influence of electromagnetic radiation. The absorption of a photon increases the vibrational amplitude by one unit and changes the vibrational quantum number from 0 (the ground state) to 1. Because the wavelengths of the pertinent photons range from 2 to 50 μm, the infra-red spectrum is also known as the *vibration spectrum*.

In principle, each absorption band in the infra-red spectrum corresponds to an amplitude change in one vibrational mode of the molecule. Now, for a molecule of N atoms the number of different vibrational modes is equal to $3N-6$ (or $3N-5$ for a linear molecule such as CO_2). Also, the spectral width of an absorption band varies from 0·1 to 0·5 μm as a result of collisional broadening and the limited resolving power of the spectrometer (Fig. 5.6). Consequently, it might be expected that the great number of infra-red absorption bands predicted for larger molecules overlap strongly and gradually blend together. The visual appearance of an infra-red spectrum might then be expected to resemble that of an ultra-violet spectrum, and consist of broad wavy bands.

Actually, the usual infra-red spectrum is much simpler. Even for large molecules the number of strong absorption bands rarely exceeds a value of ten (compare Fig. 5.2b). One reason for the limited number of absorption bands is the fact that not every molecular vibration can change its amplitude under the influence of electromagnetic radiation. As has been remarked in § 2.4, only those vibrations can be excited by electromagnetic radiation for which the total dipole moment of the molecule changes during the vibration. The vibration is then said to be *infra-red-active*.

As an example we consider the linear molecule CO_2 with a total dipole moment equal to zero in the equilibrium configuration (Fig. 5.5). A possible mode of vibration is the periodic, simultaneous, opposite displacement of the oxygen atoms with respect to the carbon atom, which remains at rest. During this *symmetric stretching vibration* the dipole moment remains zero. Accordingly, this vibration is infra-red-inactive and does not produce an absorption band in the infra-red spectrum. However, if the oxygen atoms oscillate perpendicularly to the molecular axis and the carbon atom moves in the opposite direction to keep the

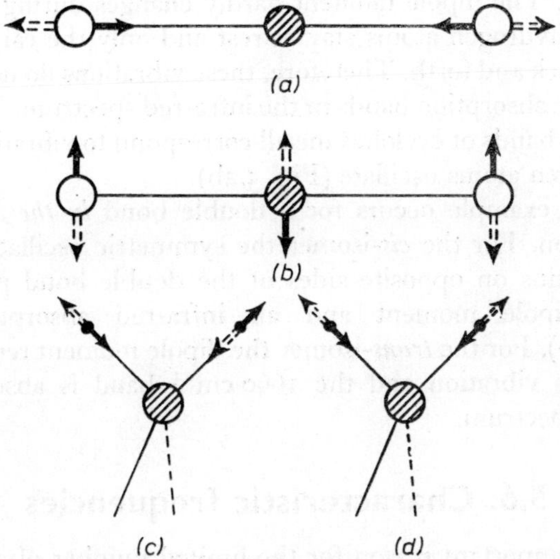

FIG. 5.5. Vibrational modes of simple atomic groups. (a) Symmetric stretching vibration of CO_2; infra-red-inactive. (b) Bending vibration of CO_2; infra-red-active. (c) Antisymmetric stretching vibration of the CH_2-group; infra-red-active. (d) Symmetric stretching vibration of the CH_2-group; infra-red-active.

centre of mass of the molecule at rest, the dipole moment varies periodically during the vibration. Consequently, the *bending vibration* of CO_2 is infra-red-active and yields an absorption band at 667 cm^{-1}.

The symmetric vibration of molecules such as CCl_4 provides a second example of an infra-red-inactive vibration mode. When all four chlorine atoms oscillate back and forth along the C–Cl bonds, the dipole moment remains permanently zero and no infra-red absorption occurs (this is known as the ***breathing vibration*** of CCl_4).

Now, larger molecules rarely show the high symmetry of CO_2 or CCl_4 and, strictly speaking, for none of their vibrations is the spectral transition absolutely 'forbidden' for symmetry reasons. However, the above considerations are helpful in understanding why some vibrations of larger molecules give rise only to very weak absorption bands.

The dipole moment of saturated, non-chained hydrocarbons is very small. The dipole moment hardly changes during vibrations when the hydrogen atoms stay at rest and only the carbon atoms oscillate back and forth. Therefore, these vibrations do not produce measurable absorption bands in the infra-red spectrum. The strong absorption bands of cyclohexane all correspond to vibrations where the hydrogen atoms oscillate (Fig. 5.2b).

Another example occurs for a double bond ***in the centre*** of a hydrocarbon. For the *cis*-isomer the symmetric oscillation of the carbon chains on opposite sides of the double bond produces a varying dipole moment and an infra-red absorption band (1660 cm^{-1}). For the ***trans***-isomer the dipole moment remains zero during this vibration and the 1660 cm^{-1} band is absent in the infra-red spectrum.

5.6. Characteristic frequencies

The most important reason for the limited number of absorption bands in the infra-red spectrum is the fact that the frequencies of many vibrational modes are equal and together give rise to a single absorption band in the spectrum. In a hydrocarbon molecule the carbon atoms are so much heavier than the hydrogen atoms that they are hardly influenced by the oscillations of hydrogen atoms only. Therefore, the oscillations of two hydrogen atoms far apart in the molecule are practically mutually independent. If, in addition, both atoms are chemically similar, for example when each atom is part of a CH_3-group in the molecule, the vibrational modes and the vibrational frequencies will be virtually equal and give rise to a single absorption band in the spectrum. The absorption band at 2900 cm^{-1} in the spectrum of cyclohexane results from the oscillation of hydrogen atoms along their bonds with the nearest carbon atom. The band is split into two peaks because the two hydrogen atoms in a CH_2-group may either move simultaneously towards and away from the carbon atom (symmetric vibration) or move in opposite directions (antisymmetric vibration) (Fig. 5.5).

In this way a number of infra-red absorption bands can be attributed to the oscillations of hydrogen atoms relative to the heavy atom with which they are bonded. To a first approximation, the position of these bands in the spectrum is *independent* of the structure of the entire molecule, but is characteristic of the hydrogen-containing group in the molecule. Accordingly, these are called *characteristic frequencies*. The spectrum of cyclohexane shows two strong absorption bands that are both characteristic of the CH_2-group in this molecule. The *stretching vibrations* of the CH_2-group are located at 2900 cm^{-1}, and the *bending vibrations* at about 1500 cm^{-1}.

In a completely similar way the hydrogen atoms bonded to nitrogen atoms or to oxygen atoms give rise to absorption bands that are characteristic of the presence of N–H or O–H bonds in the molecule. Thus, amines show a weak characteristic absorption band at around 3300 cm^{-1}, and alcohols show a characteristic band at 3600 cm^{-1} (provided that low concentrations are measured).

The phenomenon of characteristic frequencies is not restricted to hydrogen atoms. Heavier diatomic groups, such as $C{\equiv}N$ and $C{=}O$, also show distinctive absorption bands in the infra-red spectrum. Larger groups, such as aromatic rings, can also be recognized by their characteristic vibrational frequencies.

This does not imply that, conversely, each absorption band of an infra-red spectrum can be ascribed to the vibration of well-defined atomic groups in the molecule. It should be emphasized that the concept of characteristic frequencies is an empirical approximation, although an extremely valuable one. In principle, however, all atoms of the molecule oscillate during a vibration, and strictly speaking a group of atoms cannot be considered as completely isolated from the remainder of the molecule. As a result, the exact location of an absorption band varies with the structure of the molecule, even in the case of a characteristic frequency. In fact, the minor variations of the position of the carbonyl stretching vibration around 1700 cm^{-1} can be used to distinguish between aldehydes and ketones or between different esters.

Accordingly, the characteristic frequency is expected to be found within a certain wavenumber range rather than at a fixed point in the spectrum. In favourable cases (stretching vibrations) the variation observed for different molecules is restricted to 50 cm^{-1}. In less favourable cases (bending vibrations) the variation may be more than 100 cm^{-1}. If the range becomes too large, it is no longer justified to speak of a characteristic frequency. The absorption band cannot even approximately be ascribed to the oscillation of a few isolated atoms in the molecule, but is due to the vibration of all atoms and is characteristic of the molecule as a whole, that is of the substance in question.

5.7. Survey of the infra-red spectrum

From the examination of a large number of infra-red spectra, supported by theoretical calculations, some general correlations have been derived between the nature of a molecular vibration and the position of the corresponding absorption band in the spectrum. In doing this, the concepts of force constant and reduced mass, which have been discussed for diatomic molecules in § 2.4, are extended to larger molecules.

(*a*) The 0–1 vibrational frequencies (*fundamentals*) range from 4000 cm^{-1} to 200 cm^{-1}. In addition, *overtone bands* and *combination bands* may be observed, but these bands are very much weaker than the corresponding fundamentals. Overtone bands arise from 0–2 transitions of the vibrational quantum number, and combination bands are due to the simultaneous excitation of two molecular vibrations.

(*b*) The force constant of a stretching vibration (periodic change of bond lengths) is appreciably larger than the force constant of a bending vibration (periodic change of bond angles) for the same group. Absorption bands from stretching vibrations are therefore found at larger wavenumbers.

(*c*) The wavenumber of stretching vibrations of diatomic groups decreases with increasing mass of the oscillating atoms.

(*d*) The force constant of a triple bond is larger than that of a double bond and, in turn, this is larger than the force constant of a single bond. The absorption bands of the stretching vibrations of these groups are observed at correspondingly larger wavenumbers. (The force constant is not a direct indication of the *chemical* strength of a bond. Although the atoms in a double bond are tightly connected, the double bond is easily broken in chemical reactions, because new single bonds are formed with a resulting net energy gain for the reaction.)

The above considerations are substantiated by the following sub-division of the infra-red spectrum.

(a) 10000–4000 cm⁻¹ (1–2·5 μm)

This *near* infra-red region can best be studied with a spectrometer for the visible region (tungsten lamp, silica cells, photoconductive detector). Because the (weak) absorptions result exclusively from overtone bands of hydrogen stretching vibrations, this region is of limited analytical interest.

(b) 4000–1600 cm⁻¹ (2·5–6 μm)

Nearly all absorption bands in this part of the spectrum can be ascribed to the stretching vibrations of specific atomic groups in the molecule. Accordingly, this is called the region of the characteristic frequencies. It should be emphasized, however, that the phenomenon of characteristic frequencies is not restricted to this region. Further subdivision yields:

(i) *4000–2500 cm⁻¹ (2·5–4 μm)* Absorption from stretching vibrations of hydrogen-containing groups, such as —OH, $—NH_2$, ≡CH, $>CH_2$, $—CH_3$.

(ii) *2500–2000 cm⁻¹ (4–5 μm)* Absorption from triple bonds (C≡N, C≡C) and allenes (C=C=C).

(iii) *2000–1600 cm⁻¹ (5–6 μm)* Absorption from stretching vibrations of double bonds, such as carbonyl (C=O) and alkene (C=C).

(c) 1700–650 cm⁻¹ (6–15 μm)

Traditionally, the analytically important part of the infra-red spectrum has been restricted on the low wavenumber side by the transparency limit of the sodium chloride used for prisms and cell windows. The low wavenumber portion of this region presents a complicated pattern of overlapping absorption bands arising from stretching and bending vibrations of atoms connected by single bonds (C–H and O–H bending, C–C, C–O, C–N bending and stretching vibrations). Some bands are characteristic of larger units in the molecule, such as aromatic rings. In general, however, the bands cannot be related to the presence of specific structural features of the molecule. All atoms in the molecule are involved in the vibration and the shape features of this part of the spectrum are determined by the size and the structure of the molecule as a whole. Since it is characteristic for the molecule as such, this part of the infra-red spectrum is colloquially known as the *fingerprint* region.

(d) Below 650 cm^{-1} *(*$> 15\ \mu m$*)*
With modern grating instruments the historic border line of 650 cm^{-1} has become rather arbitrary, and interest in the lower wavenumber range is rapidly growing. Cell windows of KBr or CsBr permit extensions down to 400 and 250 cm^{-1} respectively, and allow the measurement of absorption bands from heavier atoms (bromides, iodides, phosphides). This part of the spectrum is also of interest in the determination of the spatial configuration of polymers. The study of lattice vibrations and pure rotation spectra in the far infra-red (< 200 cm^{-1}) has long been complicated by severe instrumentation problems (sources of radiation, detectors). To a large extent these problems have been overcome by the development of a completely different type of spectrometer, known as the Fourier spectrometer, that allows measurements from 1000 cm^{-1} down to 50 cm^{-1}. As a result, the analytical application of this region will gain importance in the next few years.*

5.8. Qualitative analysis

The organic research chemist is faced with two major analytical problems: the unravelling of complex mixtures frequently of

* Another approach to the small wavenumber region is offered by laser-Raman spectrometry, which makes possible the study of vibrational and rotational transitions down to 10 cm^{-1}. The great simplification due to the introduction of laser sources has strongly revived interest in Raman spectrometry.

similar substances, and the qualitative and quantitative assessment of the separated components. The former problem is solved mainly through chromatographic methods, whereas spectrometric tools are invaluable for the latter. Infra-red spectrometry finds its main application in the qualitative step, where it is used either for the *identification* of an unknown but traceable substance or for the elucidation of the *structure* of a newly synthesized compound.

Many compounds possess the same refractive index or the same melting point, at least within the precision of measuring these constants. On the other hand, it would be difficult to find two compounds with identical infra-red spectra. Consequently, the spectrum as a whole provides a highly characteristic fingerprint of a molecule and is very suitable for identification purposes, at least in principle. Unfortunately, the practical realization is not so simple.

In order to identify a substance by its infra-red spectrum, the spectrum must be compared with reference spectra of known compounds. This is comparatively easy, when there is a strong indication what the substance might be. In that case the spectra of the unknown substance and the reference compound are recorded and mutually compared. This procedure becomes time-consuming when the number of possible alternatives is ten or more, and it is completely inapplicable if there is no prior indication at all or if the reference substance is not available.

One obvious solution is to compare the spectrum of the unknown with previously recorded spectra. If, as is often the case, a laboratory is working in a particular field (for example fatty acids), it may collect its own library of reference spectra. Alternatively, we can resort to one of the commercially available collections of infra-red spectra run under standard conditions and reproduced or coded on index cards. It is obviously impossible to compare these one by one with the spectrum of the unknown substance, and various ingenious selection procedures have been devised to find the correct spectrum as quickly as possible. All search procedures start by matching a few characteristic strong absorption bands whose presence (or absence!) is noted in the spectrum of the unknown substance. Naturally, additional information such as melting point, refractive index, elementary composition etc. will be extremely helpful. Still, it is not an easy task to isolate the correct spectrum from a collection of a few thousand spectra. The advent of high-speed digital computers has stimulated many attempts to describe infra-red spectra by a series of numbers so that the selection can be performed by the computer. So far, however, these attempts have not been entirely successful.

One obvious limitation of all search procedures is the unavoidable incompleteness of any library of infra-red spectra. This is experienced most acutely by scientists synthesizing novel compounds. The chemist will then try to deduce the structure of the unknown compound from an *interpretation* of its combined infra-red, nuclear magnetic resonance and mass spectra.

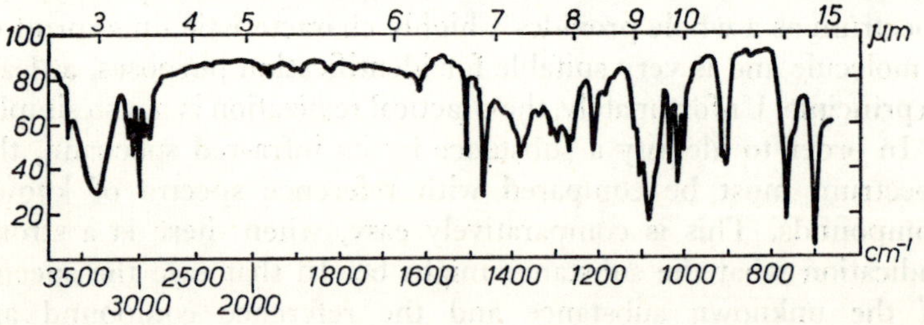

FIG. 5.6. Infra-red spectrum of an unknown substance, discussed in the text.

With the perfection of the latter two methods the contribution of infra-red spectrometry to structural analysis has markedly decreased over the past few years, but still the presence of particular absorption bands in the infra-red spectrum provides rapid confirmation of the presence of certain functional groups (through the characteristic frequencies). The exact location of these bands also provides information about the position of the functional groups in the molecule.

As an example we consider the spectrum shown in Fig. 5.6. With the assistance of correlation charts and tables found in textbooks on infra-red spectrometry, the following conclusions can be drawn (compare Fig. 5.7).

The small sharp peak at 3600 cm^{-1} indicates an OH stretching vibration. Because molecules containing this group tend to associate by formation of hydrogen bridges, the frequency of this vibration may be decreased, which results in the broad band around 3400 cm^{-1}.

The composite band at 3000 cm^{-1} is due to CH stretching vibrations. Peaks above 3000 cm^{-1} indicate the presence of unsaturated carbon bonds, those below 3000 cm^{-1} the presence of saturated carbon bonds.

Strong bands at 1450 cm^{-1} result from C=C stretching vibrations, presumably in an aromatic ring. This is also indicated by the absorption bands at 760 and 700 cm^{-1}. The band at 760 cm^{-1} points to a mono-substituted aromatic ring, which is confirmed by the pattern of weak absorption bands between 2000 and 1700 cm^{-1}. These are combination bands (simultaneous change of two vibrational frequencies), which may be better observed in the spectrum of a more concentrated solution of the substance. Finally, the band at 1080 cm^{-1} indicates a C–O bond, probably from a primary alcohol.

From this information we conclude that the molecule contains a mono-substituted aromatic ring, a saturated aliphatic chain and a primary alcohol group. A possible structure is C_6H_5–R–OH, where R is still unknown. A definite conclusion may be reached with the aid of other data, such as molecular mass, elementary composition etc. If the spectrum of Fig. 5.6 is compared to other spectra selected on the basis of the above information, it would be found that the substance is 2-phenylethanol ($C_6H_5CH_2CH_2OH$).

This example demonstrates that infra-red spectrometry alone is generally insufficient to derive the structure of an unknown compound. Especially in more complicated cases (steroids, pesticides, proteins) the more powerful techniques of nuclear magnetic resonance spectrometry (Chapter 10) and mass spectrometry (Chapter 11) are required. It is also clear that not all absorption bands in the spectrum can be interpreted. Many moderately strong bands in the fingerprint region cannot be attributed to particular atomic groups.

The foregoing discussion shows that, for the identification as well as for the interpretation of an infra-red spectrum, purity of the unknown substance is an essential requirement. (It is possible, however, to detect certain impurities in known substances.) Therefore, preliminary separations form an important step in qualitative organic analysis. As was remarked above, chromatographic methods are frequently applied, but the amount of sample often creates a problem. In gas chromatography and thin-layer chromatography optimum separations are achieved when the total amount of sample is restricted to a few microlitres. The amount recovered of a single component will then be of the order of a few micrograms. This is much less than the milligram normally used to run an infra-red spectrum. Special techniques have been developed to obtain adequate spectra from as little as one to ten micrograms of sample, which is sufficient for components separated by thin-layer chromatography. Alternatively, preparative gas chromatographs yield much larger amounts of substance at the cost of somewhat poorer separations. On the other hand, the powerful separation method of capillary-column gas chromatography yields only about 0·01 μg per separated component. Such small amounts can only be handled by mass spectrometers (Chapter 11).

5.9. Quantitative analysis

Although the Beer–Lambert law (Equation 4.2) is also valid for infra-red absorption bands, the quantitative applications of infra-red spectrometry are rather limited for several reasons. Because

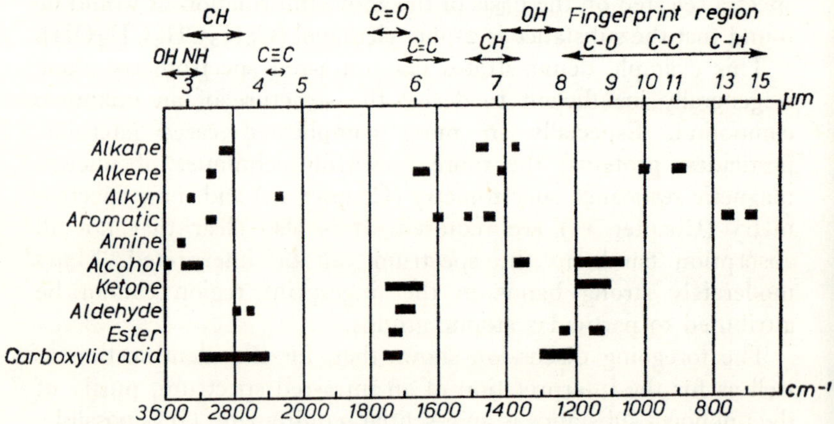

FIG. 5.7. Characteristic frequencies in the infra-red spectrum. For each type of compound only those absorption bands have been indicated that are characteristic for that particular functional group. Naturally, an alcohol, such as butanol, will show the characteristic hydroxyl bands as well as the alkane bands of the CH_3 and CH_2 groups.

vibrational absorption bands are relatively sharp, the apparent band shape depends strongly upon the operating conditions of the spectrometer (notably upon the slit width; compare Fig. 4.6). Analytical curves based on band maxima are frequently bent. The use of band areas gives better results, but it is often difficult to determine the boundaries of an absorption band and to correct for unavoidable background absorption from the solvent. Also, the quantitative precision of the optical-null system utilized in infra-red spectrometers (§ 5.1) is only fair.

As a result, infra-red spectrometry is generally inferior to either ultra-violet spectrometry for the determination of a single compound, or gas chromatography for the quantitative analysis of mixtures of organic substances. Therefore, the quantitative application of infra-red spectrometry is restricted to those cases where other methods of analysis fail. A typical example is the measurement of OH absorption bands at 3400 cm^{-1} for the determination of the number of hydroxyl groups in polymers, from which the length of the polymer chain can be deduced. Perhaps the most important quantitative application of infra-red spectrometry is in the field of gas analysis. For example, the concentration of carbon monoxide or carbon dioxide in air can be determined and monitored by measuring the absorption bands at 2150 cm^{-1} (Fig. 2.4) or 2350 cm^{-1}, respectively, with a simple, single-beam filter instrument.

5.10. Inorganic applications

The application of infra-red spectrometry to inorganic analysis is a fairly recent and largely unexplored extension of the method. Compound ions, such as sulphate, phosphate and silicate give characteristic infra-red absorption bands, although it must be admitted that the bands are rather broad and void of marked structure. Since water is an extremely inconvenient solvent for infra-red spectrometry, solution analysis is generally impossible. However, the KBr-pellet technique (§ 5.4) has been successfully applied to the qualitative (and occasionally the quantitative) analysis of minerals and glasses.

Bibliography

1. BAUMAN, R. P., *Absorption Spectroscopy*, Wiley, New York, 1962.
2. BELLAMY, L. J., *The Infra-red Spectra of Complex Molecules*, Wiley, New York, 1958.
3. CONLEY, R. T., *Infra-red Spectroscopy*, Allyn and Bacon, Boston, 1966.
4. CROSS, A. D., *An Introduction to Practical Infra-red Spectroscopy*, Butterworths, London, 1960.
5. NAHANISHI, K., *Infra-red Absorption Spectroscopy—Practical*, Holden Day, San Francisco, 1962.
6. BELLAMY, L. J., *Advances in Infra-red Group Frequencies*, Methuen, London, 1968.
7. SILVERSTEIN, R. M., and BASSLER, G. C., *Spectrometric Identification of Organic Compounds*, Wiley, New York, 2nd ed., 1967.
8. WILLIAMS, D. H., and FLEMING, I., *Spectroscopic Methods in Organic Chemistry*, McGraw Hill, New York, 1966.
9. HILL, D. W., and POWELL, T., *Non-dispersive Infra-red Gas Analysis in Science, Medicine and Industry*, Adam Hilger, London, 1968.

Problems

1. A striking feature of the optical part of an infra-red spectrometer is the large number of plane and concave mirrors used. Why are mirrors preferred to lenses?

2. In a double-beam ultra-violet spectrometer the slit width is controlled by the magnitude of the reference signal (§ 4.2). If this principle were to be utilized in an infra-red spectrometer, what would happen at wavelengths of solvent absorption, and how would the resolving power at those wavelengths be influenced?

3. Predict the wavenumbers of strong absorption bands of acetone. Will this be a suitable solvent in infra-red spectrometry?

4. Will the C≡C stretching vibration of acetylene (C_2H_2) show an absorption band in the infra-red spectrum? What conclusion may be drawn about the C≡C vibration in normal alkynes?

5. Why is the spectrum in Fig. 5.6 not due to a ketone?

6. A weak absorption band at 2·94 μm is suspected to arise from an overtone band. What will be the wavenumber of the corresponding fundamental band?

7. The spectrum of specially purified KBr shows a broad absorption band at 3400–3200 cm^{-1}, which disappears when the KBr is kept at 150°C for some time. What does this indicate?

8. Compare the spectra of fluorolube and carbon tetrachloride, and explain why no absorption bands from C–Br and C–I vibrations are observed between 4000 and 650 cm^{-1}.

9. Equation 4.2 does not apply to a pure substance. Can you alter the derivation in § 4.5 in such a way that an expression of the Beer–Lambert law results which is applicable to pure materials?

10. Indicate the wavelength ranges that are transmitted by each of the four filters used to prevent order overlap in the infra-red spectrometer of Fig. 5.1a.

11. Study the pictures of Fig. 5.5 carefully. Could you suggest the oscillation pattern for the antisymmetric stretching vibration of CO_2? Will this vibration be infra-red-active?

CHAPTER SIX

FLAME SPECTROMETRY

6.1. Flames and burners

Flame spectrometry is concerned with the interaction between electromagnetic radiation and free atoms generated in a chemical flame. A typical example of the flame and burner system used is shown in Fig. 6.1. A gas tank supplies a regular flow of a combustible *fuel gas*, such as acetylene. High-pressure air enters the

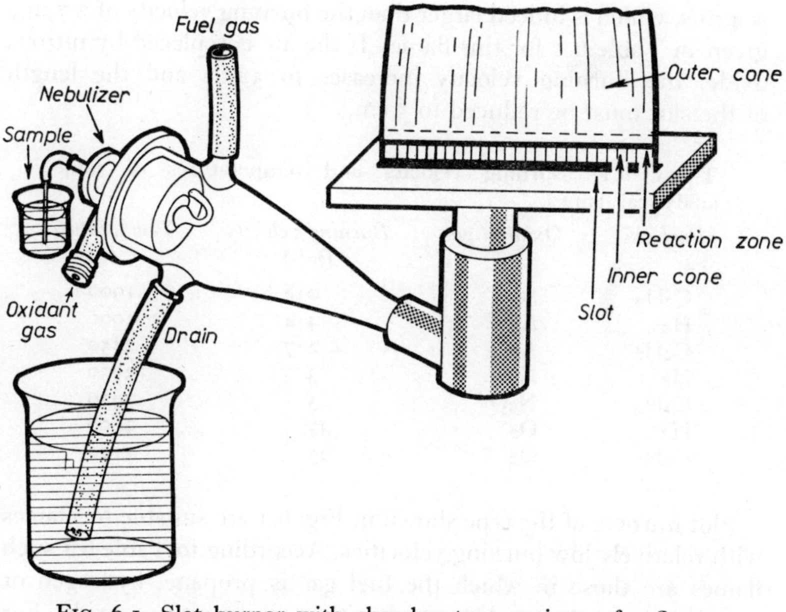

FIG. 6.1. Slot burner with chamber-type aspirator for flames with low burning velocity.

burner through another tube. Fuel and oxidant gas are thoroughly mixed in the spray chamber and leave the burner through a slot 10 cm long and 0·05 cm wide.

The flame is ignited by heating the gas locally with a lighted match to such a degree that a chemical reaction is initiated between acetylene and oxygen from the air. The energy released in this exothermic reaction heats the molecules in the immediate vicinity to their *ignition temperature*, and in this way the reaction proceeds with a finite *burning velocity* towards the burner.

In order to prevent the reaction from reaching the burner (which would cause the flame to flash back) the gas mixture is made to flow very rapidly through the burner slot. If the linear flow velocity of the flame gases is somewhat larger than the burning velocity, the combustion reaction remains localized within a thin layer above the burner called the *primary reaction zone*. This condition puts an upper limit to the dimensions of the burner slot. For example, for a flame burning on 2 l/min of acetylene and 10 l/min of air the linear flow velocity through a $10 \times 0{\cdot}05\ cm^2$ slot is 4 m/s, which is indeed larger than the burning velocity of 2·7 m/s given in Table 6.1 for this flame. If the air is replaced by nitrous oxide, the burning velocity increases to 5 m/s and the length of the slot must be reduced to 5 cm.

TABLE 6.1. Burning velocity and temperatures of some analytical flames

Fuel gas	*Oxidant gas*	*Burning velocity (m/s)*	*Temperature (°K)*
C_3H_8	Air	0·8	1900
H_2	Air	4·4	2000
C_2H_2	Air	2·7	2450
H_2	N_2O	3	2850
C_2H_2	N_2O	5	2950
H_2	O_2	37	2800
C_2H_2	O_2	25	3100

Slot burners of the type shown in Fig. 6.1 are suitable for flames with relatively low burning velocities. According to Table 6.1 such flames are those in which the fuel gas is propane, hydrogen or acetylene, and the oxidant gas is either air or nitrous oxide. For reasons to be discussed in § 6.3 and 6.4 the slot geometry is preferred for flame emission and flame absorption spectrometry, but alternatively these flames may be supported on Meker-type burner heads with a number of circular holes arranged in concentric rings. In all cases the total opening is relatively large (0·2–0·5 cm^2) and the gas mixture leaves the burner with a moderate velocity ($<$ 10 m/s) in parallel streams. This is called a *laminar*, *premixed* flame.

The chemical energy produced by combustion heats the flame gases until a uniform, stable temperature is established. This temperature is predominantly determined by the composition of the flame and forms a characteristic property of a particular combination of fuel and oxidant gases (Table 6.1). Maximum temperatures are reached in *stoichiometric* flames, where the supply of oxidant gas is just sufficient for the complete combustion of the fuel gas. In a stoichiometric hydrogen–oxygen flame the flow rate of hydrogen gas is twice the flow rate of oxygen gas. In practice, *fuel-rich* flames are preferred, i.e. those containing an excess of fuel gas (§ 6.2). The temperatures presented in Table 6.1 refer to fuel-rich flames, and are 50–100 degrees lower than the maximum possible values. It is seen that the temperatures of analytically useful chemical flames cover a range from 2000 to 3000°K. In flames supported by air a large amount of nitrogen must be heated, and this restricts the flame temperature to a maximum of about 2500°K. High-temperature laminar flames can be obtained when air is replaced by nitrous oxide.

The figures in Table 6.1 show that high flame temperatures can also be reached if pure oxygen gas is used instead of nitrous oxide, which would be advantageous from an economic point of view. However, this also raises the burning velocity of the flame tenfold and, in turn, the exit opening of the burner must be reduced to a few mm^2. A suitable burner is shown in Fig. 6.2. The fuel gas (hydrogen or acetylene) and the oxidant gas (oxygen) flow through concentric tubes and enter the flame separately. The gases are mixed in the flame, partly by diffusion and partly by the turbulences resulting from the high linear flow velocity (100 m/s). Accordingly, this is called a *turbulent diffusion* flame.

Fig. 6.2 also illustrates the principle of *pneumatic aspiration* of the sample solution. The rapid flow of support gas produces a pressure drop at the upper end of a capillary tube sticking into the sample

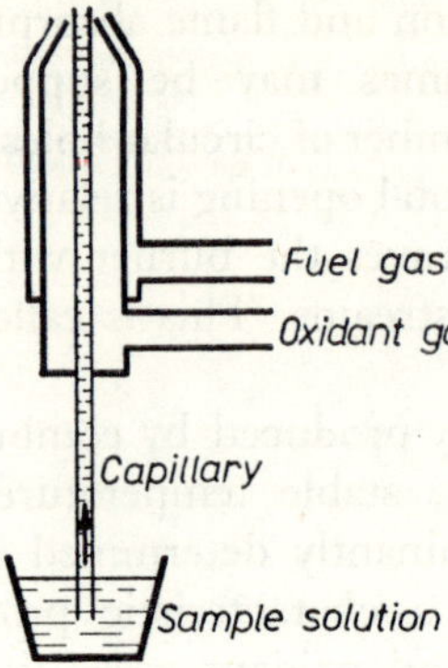

FIG. 6.2. Total-consumption burner for flames with high burning velocity.

solution. The solution is drawn into the tube and nebulized as a spray of small droplets into the flame. Because in the arrangement of Fig. 6.2 all solution entering the capillary is aspirated into the flame, this is called a *total-consumption burner*. Pneumatic aspiration is also utilized in the burner shown in Fig. 6.1. Here, however, the sprayed droplets are partly broken up against a glass ball in front of the capillary. The larger droplets condense against the wall of the spray chamber and flow back through the drain. Only a fine mist of the smallest droplets actually reaches the flame.

Total-consumption burners have been used extensively in flame emission spectrometry and are still used occasionally in atomic fluorescence spectrometry. However, one annoying aspect of the turbulent flame is its loud noise, and the popularity of total-consumption burners has decreased rather sharply since the introduction of high-temperature nitrous-oxide-supported laminar flames.

6.2. Physical and chemical processes in the flame

Having discussed the macroscopic configuration of the burner system we must now consider the microscopic behaviour of particles and molecules inside the flame. Immediately after leaving the burner the flame gases pass an inner cone where they are heated to their ignition temperature (600–800°K). In the adjacent reaction zone of a few tenths of a millimetre, a rapid succession of chemical reactions occur, mainly through the action of radicals (OH, CH). The released energy heats the gas, and a maximum temperature is reached slightly above the reaction zone. In the outer cone the radicals recombine to form molecules and the temperature gradually decreases. In a turbulent flame the various zones are not as clearly distinguished as in a laminar flame. The turbulent flow draws ambient air into the flame and the combustion penetrates deeply within the outer cone.

From an analytical point of view the primary function of the flame is to produce an atomic vapour and to convert the dissolved sample molecules into free atoms at a constant rate and with optimum efficiency. Thus the nebulized droplets first release the solvent, then the remaining solid particles evaporate, and finally the sample molecules dissociate into free atoms, which are the object of flame spectrometric analysis. It should be appreciated that this sequence of events must be completed in less than a millisecond for the atoms to be formed within one centimetre from the burner head.

Whereas solvent evaporation proceeds very rapidly in all flames, sample evaporation and molecular dissociation depends upon the size of the droplets entering the flame, upon the flame temperature, and upon the chemical composition of the flame. Therefore, these factors must be optimized and kept constant. Careful regulation of the flow rates of the flame gases provides a constant flame composition and a constant flame temperature. It also ensures a regular supply of sample solution to the burner. Common flow rates are 10 l/min of flame gases and 5 ml/min of solution into the burner. In a total-consumption burner all droplets reach the flame, but the larger ones (diameter 10 μm or more) yield correspondingly larger solid particles which do not completely volatilize during their rapid flight through the turbulent flame. By contrast, in chamber-type aspirators (Fig. 6.1) the larger droplets flow back down the drain and only some 10 per cent of the aspirated solution reaches the flame. However, this fraction consists only of tiny droplets (diameter less than 5 μm), and the resulting solid particles volatilize rapidly. Ultimately, therefore, the efficiency of evaporation is higher in premixed, laminar flames.

Obviously, the volatilization of solid sample particles and the dissociation of the molecules also depend upon the characteristic properties of the sample, notably upon the heat of evaporation and upon the dissociation energy. Salts from zinc, copper, manganese and magnesium dissociate completely into free atoms even at low flame temperatures. The atomization of aluminium, silicon, titanium and tungsten is still so slight at 2500°K that these elements cannot be detected in flames supported by air. The dissociation is also promoted by reducing conditions in the flame (reduction of oxides). This explains the widespread use of *fuel-rich* flames, specifically the acetylene–air flame for easily atomized elements and the acetylene–nitrous oxide flame for elements that atomize only with difficulty.

Incomplete atomization is also a major source of interferences in flame spectrometric analysis. For example, the presence of phosphate in the sample disturbs the determination of calcium through the formation of a stable calcium phosphate that resists dissociation. Such *chemical interferences* are a rare phenomenon in high-temperature flames. On the other hand, atoms with low ionization energy (§ 2.1), such as Na, K, Ba and Sr, tend to lose an electron in high-temperature flames. Because the analysis is based upon free neutral atoms, this ionization is undesirable. It can be effectively suppressed by the addition of excess caesium to the sample solution.

In conclusion, it may be stated that as a result of evaporation and dissociation the dissolved sample molecules are partly converted into free atoms. The vapour concentration n of free atoms of an element in the flame is proportional to the concentration c of that element in the sample solution:

$$n = Kc \tag{6.1}$$

The major advantage of flame spectrometry is the reproducibility and constancy of the proportionality constant K.

The order of magnitude of the vapour concentration is readily derived. If 5 ml of a 10^{-5}M solution is aspirated into the burner per minute, and 10 per cent of this reaches the flame, the rate of entry of the element is 3.10^{15} atoms/min. Simultaneously, 10 litres of flame gases are supplied per minute, which expand tenfold at the flame temperature (10^5 cm^3/min). Assuming complete volatilization and dissociation, the vapour concentration of free atoms of the element in the flame is then found to be $n = 3.10^{10}$ atoms/cm³.

6.3. Atomic emission flame spectrometry

The atoms and molecules in a flame are in perpetual motion. Through collisions kinetic energy can be transferred to a valence electron in an outer shell of an atom, which is then raised to a higher orbit. This raises the potential energy of the atom. Within a very short time a stable *dynamic* equilibrium is reached, where the number of atoms gaining potential energy in collisions is exactly equal to the number of excited atoms losing their potential energy. According to the Boltzmann equation given in § 1.5, the result is

$$\frac{n_s}{n_0} = \frac{g_s}{g_0} e^{-E_s/kT} \tag{6.2}$$

where n_0 is the concentration of the atoms of an element in the ground state and n_s is the concentration of atoms in a level with an excitation energy E_s; g_s and g_0 are weighting factors, k is Boltzmann's constant, and T is the flame temperature.

If the atoms are distributed over several energy levels, the total free atom concentration is given by

$$\begin{aligned} n &= n_0+n_1+n_2+\ldots \\ &= \frac{n_0}{g_0}[g_0+g_1\,e^{-E_1/kT}+g_2\,e^{-E_2/kT}+\ldots] \\ &= \frac{n_0}{g_0}B \end{aligned} \tag{6.3}$$

where the factor in brackets, denoted by B, is called the *partition function*, and is constant for a given flame temperature. Substitution of this expression into Equation 6.2 yields

$$n_s = \frac{ng_s}{B} e^{-E_s/kT} \tag{6.4}$$

and it is seen that the concentration of atoms in a particular excited energy level s is proportional to the *total* concentration of free atoms of that element in the flame.

Although most excited atoms lose their potential energy through collisions, some atoms convert this energy to electromagnetic radiation by releasing a photon. The intensity of this *spontaneous emission* is equal to

$$I = Ah\nu bn_s = \frac{Ag_s h\nu bn}{B} e^{-E_s/kT} \tag{6.5}$$

where b is the optical path length (the diameter) of the flame in the direction of the spectrometer, $h\nu$ is the photon energy, and A is the transition probability, expressed in reciprocal seconds, which determines the basic strength of a spectral line (§ 2.6). Combination of Equations 6.5 and 6.1 finally yields

$$I = \frac{Ag_s h\nu bKc}{B} e^{-E_s/kT} = K'c \tag{6.6}$$

Thus the intensity of an atomic spectral line emitted by the flame is proportional to the solution concentration c of the element. This expression forms the basis of quantitative atomic emission flame spectrometry. Radiation from the flame is directed by a lens towards a monochromator, which isolates the spectral line of interest from other lines emitted by the flame (Fig. 6.3a). As discussed in § 3.7, the intensity I is proportional to the radiative power emerging from the exit slit of the monochromator, and in turn this is converted into a proportional electrical signal by a photomultiplier (§ 4.3); this signal is amplified and measured by a galvanometer or a recorder. Consequently, the deflection of the meter or the recorder pen is proportional to the solution concentration c, and the value of the proportionality factor is readily established by an analytical curve determined with standard solutions of known concentration.

The value of the proportionality constant K' in Equation 6.6, and hence the sensitivity of the determination, increases with increasing flame temperature. Consequently, high-temperature flames (C_2H_2–O_2, C_2H_2–N_2O) are preferred in flame emission spectrometry. The exponential factor in Equation 6.6 decreases rapidly with increasing excitation energy E_s (compare Table 1.2, p. 15). Therefore, flame emission analysis is most fruitful for elements with resonance lines above 300 nm. (As was mentioned in

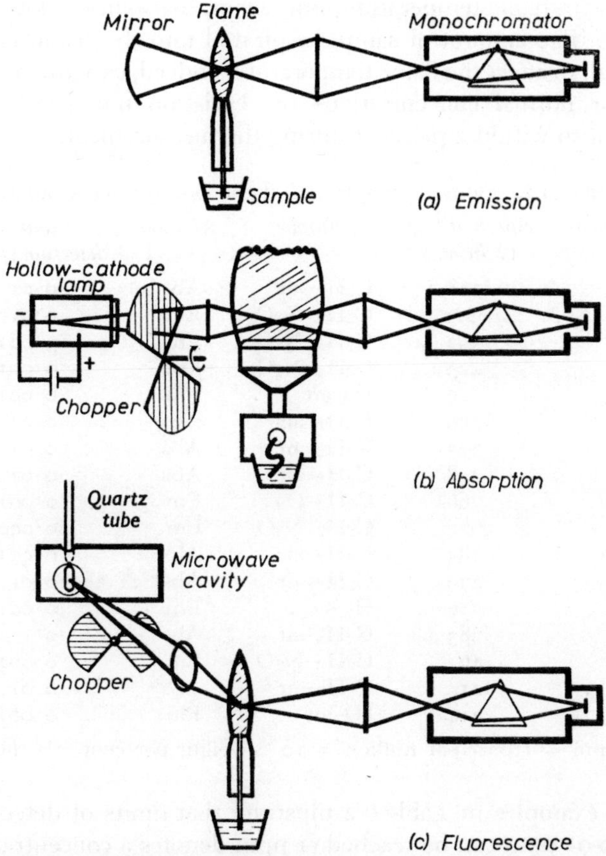

FIG. 6.3. Optical arrangements for flame spectrometric analysis according to the emission, the absorption, and the fluorescence method, respectively.

§ 2.2, a resonance line is a spectral transition for which the lower energy level is the ground state.)

It is also apparent from Equation 6.6 that the emission intensity is rather sensitive to variations in the flame temperature. As was remarked in § 6.1, the flame temperature is determined by the composition of the flame. As a result of the regular flow of the flame gases, the flame composition and the flame temperature are extremely constant. Solvent evaporation does produce a slight decrease in flame temperature, but, again, the solution flow is very regular. The amount of sample aspirated into the flame is much too small to affect the flame temperature. Indeed, experiment shows that for normal concentrations the emission intensities remain constant to within 2 per cent during the measurement.

TABLE 6.2. Some examples of flame spectrometric analysis

Element	*Spectral line* (λ *in nm*)	*Flame*	*Method*	*Limit of detection (ppm)**
Ag	328	C_2H_2–air	Abs	0·01
Al	396	C_2H_2–N_2O	Em	0·01
Ba	553	C_2H_2–N_2O	Em	0·001
Ca	423	C_2H_2–N_2O	Em	0·001
Cd	229	H_2–air	Flu	0·001
Cr	359	C_2H_2–air	Abs	0·01
Cu	325	C_2H_2–air	Abs	0·01
Fe	248	C_2H_2–air	Abs	0·01
K	766	C_2H_2–O_2	Em	0·001
Li	671	C_2H_2–N_2O	Em	0·0001
Mg	285	C_2H_2–air	Abs	0·001
Mn	279	C_2H_2–air	Abs	0·01
Na	589	H_2–O_2	Em	0·001
Pb	283	C_2H_2–air	Abs	0·1
Sr	461	C_2H_2–N_2O	Em	0·001
Zn	214	C_2H_2–air	Abs	0·01
Zn	214	H_2–air	Flu	0·001

* 1 ppm = 1 part per million = 10^{-4} weight per cent = 1 mg/l.

The examples in Table 6.2 illustrate that limits of detection of about 0·01 ppm can be reached (1 ppm denotes a concentration of 1 mg/l). Analytical curves usually cover a concentration between 1 and 100 ppm. At still higher concentrations radiation emitted by excited atoms in the flame may be reabsorbed by atoms in the ground state of that same element in the flame. This phenomenon of *self-absorption* reduces the emission intensity and, as a result, the analytical curve is bent towards the concentration axis.

6.4. Atomic absorption flame spectrometry

As is clear from Table 1.2 (p. 15), the great majority of the atoms in a flame are in the ground state even at high temperatures. In fact, elements such as cadmium and zinc with resonance lines below 230 nm are so little excited in a flame that they emit virtually no radiation. However, the ground state atoms of these elements are quite able to *absorb* radiation provided that the photon energy of this radiation corresponds exactly to the energy difference between two energy levels of the free atom. This suggests that elements such as cadmium, zinc, and magnesium may be determined by the absorption of radiation rather than by the emission of radiation. This forms the basis of an analytical method known as atomic absorption spectrometry, which has become very popular over the past ten years.

Before we discuss the instrumental aspects of this method, one word of caution seems in order. It is clear from Table 1.2 that the number of atoms in the ground state is always considerably larger than the number of excited atoms, even for elements with resonance lines above 500 nm. This does not mean, however, that the limits of detection in atomic absorption spectrometry are correspondingly lower than in atomic emission spectrometry. The measurement of absorption and of spontaneous emission differ too much to permit such a simple conclusion. In practice, it appears that so far as limits of detection are concerned flame emission spectrometry must be preferred for elements with resonance lines above 400 nm. For elements with resonance lines between 300 and 400 nm both methods give comparable results. Elements with resonance lines below 300 nm can better, or even can only, be determined by atomic absorption spectrometry (or by atomic fluorescence spectrometry; see the next section). This is illustrated by the examples in Table 6.2.

Fig. 6.4 demonstrates that a continuum source of radiation is not suitable for measuring the absorption of atomic radiation. The wavelength range selected by the monochromator from the continuum spectrum of the source (for example a tungsten filament lamp, § 4.3) is determined by the spectral bandwidth of the monochromator, and for a typical spectrometer this is of the order of 0·5 nm. However, the atoms can only absorb radiation within the narrow region of the spectral line width, which was calculated in § 2.6 to be about 0·005 nm. All radiation *within* the spectral band of the monochromator, but *outside* the line profile, passes the flame unabsorbed and reaches the detector, irrespective of the concentra-

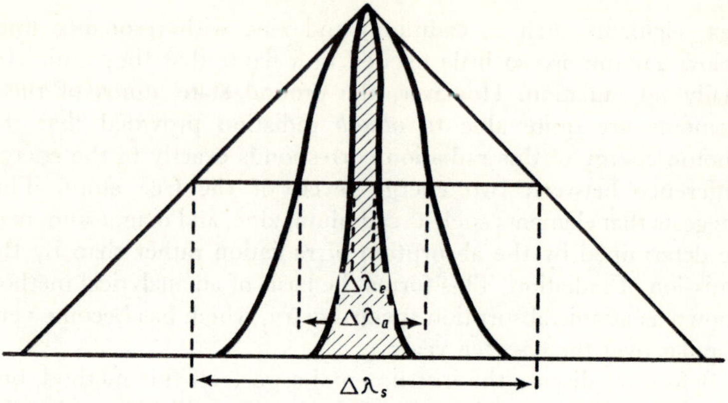

FIG. 6.4. Continuum sources and line sources in atomic absorption spectrometry.

The triangular profile represents the intensity distribution of radiation emerging from a monochromator with a spectral bandwidth equal to $\Delta\lambda_s$; $\Delta\lambda_a$ is the width of the absorption line of an element in the flame and $\Delta\lambda_e$ is the width of an atomic spectral line from the hollow-cathode lamp. The width of both line profiles has been exaggerated.

tion of absorbing atoms. Even if the absorption in the centre of the spectral band is complete, only 1 per cent of the radiation inside the spectral band will be absorbed and the *maximum* absorbance attainable is then given by $A_{max} = -\log 0{\cdot}99 = 0{\cdot}004$!

The situation is markedly improved if a source of radiation is used that emits specifically the sharp atomic spectral lines of the element to be determined. Now, the absorption line of the element in the flame is broader than the emission line of the source. This resembles the situation in molecular absorption spectrometry, where also the absorption band is broader than the band of source radiation selected by the monochromator (Chapter 4). A source of radiation meeting this specification for atomic absorption spectrometry is the *hollow-cathode discharge lamp* shown in Fig. 6.3b. A glass envelope is filled with neon at a pressure of 0·01 atm. A tungsten wire sticking through the envelope serves as the anode. The tubular hollow cathode consists of the element to be determined, for example copper. At a potential difference of 300 V and a current of 10 mA the cathode starts to glow; copper atoms are released, and are excited inside the hollow cathode by collisions with free electrons. As a result, the lamp emits the spectral lines of copper (and neon). For the observation of ultra-violet lines the lamp is provided with a silica window.

The optical system is shown in Fig. 6.3b. Radiation from the hollow-cathode lamp is focused onto the flame and from there upon the entrance slit of the monochromator, which now only serves to isolate the desired spectral line of copper from other lines emitted by the lamp. Because measurements are made at a fixed wavelength, a single-beam instrument can be used (§ 4.1). When pure solvent or a blank solution is being aspirated, the deflection of the galvanometer is adjusted to read 100 per cent transmittance (zero absorbance). Then when a copper solution is aspirated, the copper atoms in the flame absorb radiation and the recorded intensity is decreased.

One obvious consequence of using hollow-cathode lamps is that the determination of each element requires its own lamp. Most instruments have provisions for rapid interchange of lamps, and in practice this is not a serious disadvantage. However, elements such as calcium and strontium not only absorb atomic radiation, but also emit the same spectral lines in the flame. Without further provisions this would distort the absorption measurement. Therefore, a *chopper* is placed between the hollow-cathode lamp and the flame to intercept the radiation from the lamp. (Alternatively, the lamp may be run on alternating current.) If an alternating current amplifier is connected to the detector, only the periodically varying radiation from the lamp will be recorded, whereas the constant contribution from the flame will not be amplified (compare § 5.1, where the same principle is discussed for infra-red detectors). However, flame radiation does contribute to the noise (fluctuations in the signal) and, therefore, very luminous flames (C_2H_2–O_2) are better avoided.

The quantitative description of atomic absorption spectrometry is based upon the Beer–Lambert law (Equation 4.3):

$$A = -\log \frac{I}{I_0} = a'bn_0 = a'b\frac{g_0}{B}n = abn \qquad (6.7)$$

Here Equation 6.3 has been applied to convert the concentration n_0 of ground state atoms actually absorbing radiation, to the total concentration n of an element in the flame. Combination of Equation 6.7 with Equation 6.1 then yields

$$A = abKc = K''c \qquad (6.8)$$

where again K'' is a proportionality constant determined with an analytical curve using standard solutions.

In favourable cases, the *atomic absorption coefficient a* amounts to 10^{-11} cm². Hence, in a flame with an optical path length of 10 cm, vapour concentrations n of 10^{10} cm³, corresponding to a solution concentration of 10^{-5}M or about 1 ppm (§6.2) can be readily determined. Indeed, for many elements limits of detection down to 0·01 ppm can be reached (Table 6.2). As in flame emission spectrometry, the usual concentration range is 1–100 ppm. At high absorbance values ($A > 1$) the analytical curves are frequently non-linear. Probably the most important reason for this curvature is that some background radiation from the hollow-cathode lamp enters the monochromator. Note that in Fig. 6.4 radiation outside the line profile but within the spectral bandwidth of the monochromator passes the flame unabsorbed and contributes to the intensity detected by the photomultiplier. The effect is similar to the influence of stray light in molecular absorption spectrometry (§ 5.4). Another possible reason for curvature of the analytical curve is the variation of the atomic absorption coefficient over the profile of the hollow-cathode emission line. This effect is exactly similar to the variation of the molecular absorptivity over the spectral bandwidth of the monochromator discussed in § 4.5.3.

Because the absorption measurement does not require the excitation of atoms, a flame temperature of 2500°K is generally sufficient. Consequently, the acetylene–air flame is widely applied in atomic absorption spectrometry. Higher flame temperatures are definitely required to promote the dissociation of stable molecules or to remove chemical interferences (§ 6.2). For example, elements such as aluminium, titanium and silicon can only be determined in an acetylene–nitrous oxide flame.

6.5. Atomic fluorescence flame spectrometry

The excitation and the detection of atomic fluorescence radiation proceed according to principles completely similar to those considered in § 4.8 for molecular fluorescence. In Fig. 6.3c primary radiation is focused onto a flame. For geometrical reasons circular flames are preferred in atomic fluorescence spectrometry. The source of primary radiation should be very intense. Special types of hollow-cathode lamps can be utilized, but so far the best results have been obtained with *electrodeless discharge lamps*, where the energy of an electromagnetic microwave (frequency 2450 MHz, power 100 W) is transferred to a volatile compound placed in a small quartz tube. The compound, generally an iodide of the element to be determined, volatilizes and dissociates, whereupon the metallic element emits an extremely intense atomic spectrum. Again, a chopper is placed between the source of radiation and the flame so that the alternating current amplifier connected to the detector responds only to the interrupted fluorescence radiation and rejects the flame background radiation.

In analogy with Equation 4.9 the intensity of fluorescence radiation emitted by small atomic concentrations is given by

$$I_F = 2{\cdot}3\,\Phi I_0 abn \tag{6.9}$$

where I_0 is the intensity of excitation radiation from the source, Φ the fluorescence yield factor, a the atomic absorption coefficient, b the optical path length of the flame, and n the total concentration of free atoms of an element in the flame. Combination of Equations 6.9 and 6.1 yields

$$I_F = 2{\cdot}3\,\Phi\, I_0 abKc = K'''c \tag{6.10}$$

Experimental observations have confirmed that for small concentrations the fluorescence intensity is indeed proportional to the solution concentration c, and this forms the basis of quantitative analysis, in accordance with the usual analytical curve procedures.

Atomic fluorescence flame spectrometry is a fairly young method of analysis and it is still too early for a definite appraisal of its potential applicability in practice. For a few elements (zinc and cadmium; see Table 6.2) extremely low limits of detection have been reported, but it should be pointed out that these refer to ideal conditions. In order to detect the rather weak fluorescence signals, low-temperature hydrogen–air flames are frequently used to reduce the flame background noise to a minimum. Introduction of argon into the flame is essential to reduce *quenching* (§ 4.8), but even then the yield factor Φ rarely exceeds 0·1. Obviously, such low-temperature flames can give rise to serious chemical interferences and are generally unsuitable for practical analysis.

A serious limitation to the further development of atomic fluorescence spectrometry is the lack of sources of primary radiation that are sufficiently intense as well as stable. The production of electrodeless discharge lamps is still a capricious affair and the stability and reproducibility of fluorescence signals is decidedly less than in atomic absorption and atomic emission flame spectrometry. However, once this problem is solved atomic fluorescence flame spectrometry may rapidly develop into a competitive method of analysis, especially for elements with resonance lines below 300 nm.

6.6. General remarks

In the present state of the art the three flame spectrometric methods of analysis together allow the determination of more than sixty elements, including all the metals. Halogens and permanent gases (oxygen, nitrogen, noble gases) cannot be determined by flame spectrometry and the determination of phosphorus and sulphur is insensitive at best. No doubt, the widest scope is offered by atomic absorption spectrometry, but flame emission methods are gaining in popularity and, in fact, both methods should be considered as complementary. Atomic fluorescence flame spectrometry still awaits further development.

Over the usual concentration range of 1–100 ppm in aqueous solution the standard deviation of the analysis is about 0·5 per cent, which makes flame spectrometry applicable to the determination of major and minor constituents in a sample. The good precision results mainly from the high stability of the flame and the regular nebulization of the sample solution into the flame.

On the other hand, the necessity of dissolving the sample makes the analysis of ores, rocks and metal slags cumbersome and time-consuming. Also, the total salt concentration of the solution should not exceed 1 per cent, so that dissolving also implies a hundred-fold dilution. In general, this precludes the direct determination of trace constituents ($<$ 0·001 per cent, in the solid sample), and preconcentration procedures, such as solvent extraction, may be required.

A limitation of conventional flame spectrometry is the fact that rather large amounts of solution must be supplied: 1 ml of a 1 ppm solution corresponds to 1 μg of the element in a sample of perhaps 10 mg. In recent developments of the method the flame is replaced by a tiny furnace that permits the analysis of a few μl of solution, corresponding to absolute amounts of 1 ng of an element or even less.

A spectrometer of moderate resolving power (spectral band-width of 1 nm) is generally adequate for atomic absorption and atomic fluorescence spectrometry, although for optimum performance a higher resolution is to be preferred (spectral band-width less than 0·2 nm). High-resolution monochromators are essential, however, in flame emission spectrometry. It is true that, as a result of the moderate flame temperatures, the spectra of most elements show only a few strong lines (2–10), and coincidence of *atomic* spectral lines is rare. The high-resolution monochromator is required, however, to reduce *spectral interferences* due to *molecular* bands emitted by the flame itself. Notorious examples are the bands emitted by the radical CN in an acetylene–nitrous oxide flame at 420 and 380 nm (Fig. 2.1b), and the band emitted by the radical OH at 310 nm.

Some final remarks must be devoted to the subject of *inter-element effects*, by which we mean the change of the signal (intensity or absorbance) from one element due to the presence of another element in the sample.

It has been noted in § 6.2 that the presence of other cations or anions in the sample may depress or enhance the volatilization or the atomization of the element to be determined, and hence produce systematic errors. It has also been noted that such chemical interferences occur less frequently in high-temperature flames. Indeed, it is rare for inter-element effects to make a direct analysis completely impossible. Elaborate preliminary separation procedures are seldom required in flame spectrometry. Quite a number of elements can be determined without any such provisions even in low-temperature flames.

However, the analyst should always check the absence of minor inter-element effects causing systematic errors of 10–20 per cent. Most of these interferences can be effectively removed by making the composition of the standard solutions approximately equal to the composition of the sample solution, thereby ensuring that the analytical curve and the sample are measured under similar conditions. For example, in the determination of sodium in rocks, the standard sodium solutions will be supplied with calcium, aluminium and potassium.

Flame spectrometry is currently applied to the analysis of alloys, soil and agricultural products, sea water, air and water pollution, food products, blood serum etc.

Bibliography

1. HERRMANN, R., and ALKEMADE, C. T. J., *Flame Photometry*, Interscience, 2nd ed., New York, 1963.
2. MAVRODINEANU, R., and BOITEUX, M., *Flame Spectroscopy*, Wiley, New York, 1965.
3. DEAN, J. A., and RAINS, F. C. (eds), *Flame Emission and Atomic Absorption Spectrometry*, Dekker, New York, vol. I, 1969.
4. PUNGOR, E., *Flame Photometry Theory*, Van Nostrand, Toronto, 1967.
5. RAMÍREZ-MUÑOZ, J., *Atomic Absorption Spectrometry*, Elsevier, Amsterdam, 1968.
6. SLAVIN, W., *Atomic Absorption Spectroscopy*, Interscience, New York, 1968.
7. CHRISTIAN, G. D., and FELDMAN, F. J., *Atomic Absorption Spectroscopy*, Wiley, New York, 1970.

Problems

1. Check the calculation of the linear flow velocity of the acetylene–air flame in § 6.1.

2. The primary combustion reaction between acetylene and oxygen is generally assumed to yield water and carbon monoxide. Give the reaction equation, and decide whether or not a flame burning on 2 l/min acetylene and 10 l/min air is fuel-rich.

3. Equal amounts of a solution containing a constant concentration of barium chloride are aspirated into flames of increasing temperature. It is observed that the absorbance of the atomic barium line at 553 nm first increases and subsequently decreases again. Explain this phenomenon, taking into consideration that the dissociation energy of barium hydroxide is large, whereas the ionization energy of barium atoms is small.

4. Show that the excitation energy E_s in Equation 6.5 is equal to $h\nu$ for resonance lines, and explain why the resonance lines with the longest wavelength are the most intense in flame emission spectrometry.

5. Check the dimensions of the intensity expressed by Equation 6.5 with the dimensions given in § 1.1.

6. Equal amounts of sample solution are aspirated into two flames with temperatures T_1 and T_2, respectively. If evaporation and dissociation are complete in both flames, show that the intensity ratio of one and the same atomic transition in both flames is given by

$$\frac{I_1}{I_2} = \exp\left[-\frac{E_s}{k}\left(\frac{1}{T_1} - \frac{1}{T_2}\right)\right]$$

(The influence of the partition function may be ignored.) Use this expression to calculate the intensity ratio of the sodium resonance line at 589 nm in an acetylene–nitrous oxide flame and an acetylene–air flame, respectively (see also Problem 4).

7. Derive Equation 6.9.

8. The lower three energy levels of the thallium atom are at 0 (ground state), 0·96 eV and 3·18 eV. The spectral transition between the lower two levels is forbidden, but the other two transitions are allowed.

(*a*) Calculate the wavelength(s) of the atomic emission lines.

(*b*) Will the same line(s) be suitable for atomic absorption spectrometry?

(*c*) When a flame supplied with thallium is irradiated with radiation of 378 nm, two fluorescence lines are observed. If the flame is irradiated with radiation of 535 nm, no fluorescence is observed. Explain this.

9. What is the purpose of the concave mirror in Fig. 6.3a? How could this principle be applied in Fig. 6.3c?

CHAPTER SEVEN

ELECTRICAL DISCHARGES

7.1. The direct current arc

The temperatures of analytically useful chemical flames cover a range from 2000 to 3000°K (Table 6.1). Much higher temperatures can be realized with electrical discharges. The electrical sources of excitation most widely applied in emission spectrometry are the *direct current arc discharge* and the *low-voltage spark discharge*; these will both be considered in this chapter.

The electrical circuit of an arc discharge is shown in Fig. 7.1. A 250 V, 15 A direct current power source is connected through a variable load resistance R and an inductance L to two conductive electrodes, usually made of graphite. In the usual arrangement the

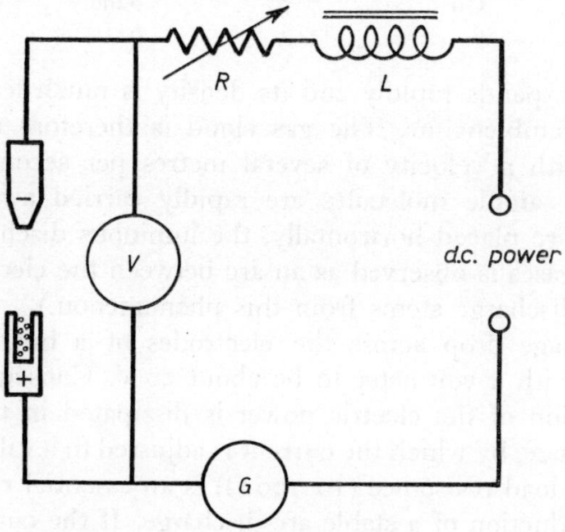

FIG. 7.1. Electrical circuit of the direct current arc.

electrodes are placed in the vertical position, free in air. Some 10–50 mg of finely powdered sample is introduced into a cavity drilled in the top of the lower electrode, which is usually the anode.

The circuit is closed by letting the electrodes touch. The electrodes are strongly heated by the electrical current, and this causes ionization of ambient air molecules. When next the electrodes are drawn apart to a separation of some 6 mm, a steady current is maintained and a stable arc discharge results. The tips of the electrodes are heated by the electrical current to a temperature of 4000°K, whereupon the carbon of the graphite electrodes sublimes and also burns with oxygen from the air. The sample melts, volatilizes, and gradually enters the discharge region; in a few minutes the entire sample is consumed. As a result of the high temperature (Table 7.1) the gaseous mixture between the

TABLE 7.1. Temperature of the direct current arc with graphite electrodes for various major constituents

Major constituent	*Ionization energy (eV)*	*Temperature (°K)*
K	4·3	5100
Li	5·4	5500
Tl	6·1	5700
Pb	7·4	5900
Al	6·9	6100
Cu	7·7	6290
C	11·3	6250

electrodes expands rapidly and its density is much less than the density of ambient air. The gas cloud is therefore transported upwards with a velocity of several metres per second, and the evaporated sample molecules are rapidly carried away. (If the electrodes are placed horizontally, the luminous discharge in the rising hot gases is observed as an arc between the electrodes; the name *arc* discharge stems from this phenomenon.)

The voltage drop across the electrodes of a burning arc is measured with a voltmeter to be about 50 V. Consequently, the major portion of the electric power is dissipated in the variable load resistance, by which the current is adjusted to a value of about 10 A. This load resistance (10–100 Ω) is an essential requirement for the production of a stable arc discharge. If the current in the circuit happens to *increase* accidentally, the potential drop across the electrodes is observed to *decrease*. Without the load resistance the current would increase even more and the system would never return to a stable equilibrium. With the circuit of Fig. 7.1, however, an accidental increase of the current still produces a decrease of the potential difference across the arc gap, but this is compensated for by the increase of the potential difference over the resistance. The inductance is included in the circuit to suppress the current fluctuations to acceptable values (0·1 A).

7.2. Physical processes in the arc

An essential feature of any electrical discharge is the stream of electrons that by their movement towards the anode carry the electrical current across the arc gap. Electrons are first released by the brilliantly white-hot tip of the cathode, which is usually the upper electrode. These electrons are accelerated towards the anode under the influence of the potential difference across the arc gap. In this process the electric energy supplied by the current source is transformed into the kinetic energy of the electrons. However, during their flight through the discharge zone the electrons collide repeatedly with heavier particles and transfer part of their kinetic energy to these particles. Through collisions, solid particles which escape from the anode are volatilized, molecules from the air and from the sample are dissociated, and atoms and molecules are excited to higher energy levels, whereupon spontaneous emission of electro-magnetic radiation can take place (§ 1.5). The characteristic blue colour of the pure graphite arc results from the band spectrum of CN radicals at 410 nm (Plate 1b).

One final but essential process is the partial ionization of atoms and molecules. In fact, the discharge as a whole is electrically neutral and the number of free electrons is exactly equal to the number of monovalent, positive ions (twofold ionization does not occur in arc discharges). Although fresh electrons are continuously supplied at the cathode, exactly the same number is carried away at the anode. However, these are *not* the same electrons! The electrons released at the cathode are quickly slowed down by collisions and captured by positive ions present in the arc column. Consequently, new electrons must be formed and the temperature of the arc discharge must rise to a value such that enough atoms or molecules ionize to supply the electrons necessary to sustain the electric current. Within a very short time (much less than a second) after the arc has been struck, the kinetic and potential energy is distributed equally over electrons, atoms, molecules and ions. The temperature of this *plasma* results from the equilibrium between the supply and loss of free electrons.

Now, more than 99 per cent of the gaseous mixture between the electrodes consists of the molecules N_2, CO, and CN, and of the atoms O, N, and C, originating from nitrogen and oxygen in the air and from carbon evaporated from the graphite electrodes. The ionization energy required to remove an electron from these constituents (§ 2.1) is very large (11–15 eV) and, consequently, the temperature must rise to very high values to ionize these constituents sufficiently. Indeed, the temperature of the pure graphite arc is found, by measurement, to be well over 6000°K (Table 7.1).

However, the ionization energy of metallic elements evaporated from the sample is much smaller, and, accordingly, these elements become ionized at correspondingly lower temperatures. Now, it does not help if we add a fraction of a milligram of, say, sodium, because although the sodium atoms ionize completely in the arc, their number is much too small to fulfil the demand for free electrons. It so happens, however, that the *total* number of metallic atoms evaporated from a sample is about equal to the number of free electrons required. Consequently, the electrons in the arc plasma are supplied predominantly by the *major metallic constituent* of the sample, and the ionization energy of this element determines the temperature of the discharge. This is illustrated by Table 7.1, where the arc temperature is given for various major sample constituents. For elements with a high ionization energy, such as copper, zinc and carbon, the arc temperature must exceed 6000°K to produce enough free electrons. On the other hand, alkali metals ionize easily and a high concentration of sodium, potassium or lithium in the sample restricts the arc temperature to 5000–5500°K. Conversely, by adding any of these elements to our samples, we can control the discharge temperature, a procedure that will be found very useful in quantitative analysis (§ 7.5).

At this point of the discussion it may be useful to note three important differences between flames and arc discharges. First, as a result of the higher temperature, the emission spectra of an arc discharge can be observed down to the far ultra-violet region of the electromagnetic spectrum. Secondly, the flame temperature results from the combustion of the flame gases and is completely independent of the composition of the sample, but the arc temperature results from the demand for free electrons and is determined primarily by the composition of the sample introduced into the graphite electrode. And finally, the forced supply of the sample solution to a flame proceeds very regularly, whereas the free evaporation of the solid sample in an arc electrode proceeds extremely erratically.

This last point must be considered further, because it necessitates the use of a particular detection system. The irregular evaporation of the sample leads to strongly fluctuating emission intensities. If the intensity of a spectral line emitted by an arc discharge were displayed instantaneously on a recorder (as with flames), the recorder trace would show both excessive noise and strong drift (§ 1.6). Similar results are observed for the spark discharge (to be discussed in § 7.6), and in spark-source mass spectrometry (§ 11.2). With all such sources instantaneous detectors are replaced by *integrating* detectors, that collect the radiation emitted over a certain *exposure time* and indicate the resulting 'sum value'. If a limited number of fixed spectral lines is to be observed, a multiple photo-electric system may be used (§ 7.7). On the other hand, if the entire spectrum is to be recorded, as in arc discharge or spark-source mass spectrometry, *photographic* detection is required, as will be discussed in the next section.

7.3. Photographic detection

Plate 3 shows an instrument for the photographic recording of arc spectra (a spectrograph). To prevent the intense, white continuum radiation emitted by the luminous graphite electrodes from entering the spectrograph, the arc is first focused onto a limiting aperture that transmits radiation from the central part of the plasma only. Either a quartz prism (as in Plate 3) or a reflection grating may be used to separate the many lines of the spectrum, the reciprocal linear dispersion amounting to 5–15 Å/mm (§ 3.2).

In the Littrow mounting (§ 4.1), the collimator lens also serves as the camera lens, and the photographic plate is placed in its focal plane (§ 3.1). This is a glass plate (10×25 cm^2) coated with a thin layer of a suspension of silver bromide grains in gelatin (the photographic *emulsion*). Through the absorption of incident photons, the silver bromide is raised to a higher energy level, the grains are sensitized and the *latent image* is formed.

When the plate is developed with a suitable reducing agent (for example hydroquinone), only the sensitized grains are reduced to free silver, but the unexposed, non-sensitized grains are left unchanged. After developing, the excess developer and the remaining silver bromide grains are dissolved, and the image remains as a black silver residue in the thin layer of gelatin fixed upon the plate (Plate 1).

As a detector the photographic plate offers several advantages. A large portion of the spectrum can be recorded simultaneously, and a permanent record is obtained. Since the plate must be exposed for some time, the intensity of the radiation is integrated over the exposure time and the random fluctuations typical of arc discharges are averaged out and smoothed. Also, very large exposure times (up to several weeks if so desired) enable the measurement of very weak intensities, such as the light from stars. Although the developing and drying of the photographic plate take time, this should be weighed against the large amount of information collected on one plate: fifteen spectra with twenty useful spectral lines each is no exception.

The principal objection to the photographic plate is the complicated relation between the photographic image and the incident radiant intensity, which is the information needed in quantitative analysis. Let us suppose a photographic plate to be illuminated for equal periods of time with increasing amounts of radiation; this may be done with a source of constant intensity and a series of filters of known transmittance (see problem 7.2).* After developing, fixing and drying the photographic plate, the 'blackness' of a spectral line in successive spectra is indeed seen to increase with

* Since the photographic emulsion reacts differently to variations in intensity or exposure time, it is not possible to use a constant intensity and varying exposure times. As a result of this *Schwarzschildt effect* all spectra must be exposed for approximately the same time.

increasing *exposure* H, the incident intensity integrated over the (constant) exposure time. Next, each spectral line (which may be, for example, 4 mm long and 0·02 mm wide) is scanned with a $2 \times 0{\cdot}01$ mm^2 spot of white light (the instrument used is appropriately called a *micro*photometer). When the beam traverses an unexposed part of the plate, the recorded intensity is adjusted to $I_0 = 100$. When the beam traverses a black spectral line, the intensity is attenuated to I. The ratio I/I_0 is called the transmittance T; this is just like the absorption measurement in molecular or atomic absorption spectrometry (§§ 4.5 and 6.4).

When measured in this way, the transmittance T of a photographic spectral line decreases with increasing exposure H, and the relation between these two variables is as shown in Fig. 7.2a.

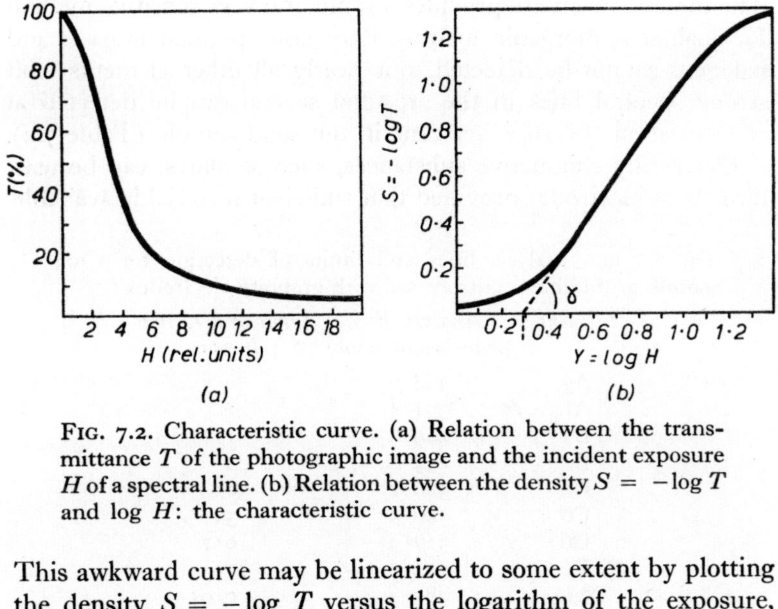

FIG. 7.2. Characteristic curve. (a) Relation between the transmittance T of the photographic image and the incident exposure H of a spectral line. (b) Relation between the density $S = -\log T$ and $\log H$: the characteristic curve.

This awkward curve may be linearized to some extent by plotting the density $S = -\log T$ versus the logarithm of the exposure, $\log H$ (Fig. 7.2b). This is called the *characteristic curve* of the photographic plate.* The shape of the characteristic curve is

* The slope of the linear portion of the characteristic curve is called the *gamma* or the contrast of the photographic emulsion. The intercept of the extrapolated linear portion on the exposure axis is called the *inertia*, and determines the speed of the emulsion.

influenced by many factors. The type of photographic emulsion, the temperature of the developer and the period of development can all be readily controlled. Unfortunately, the characteristic curve also varies with the wavelength of the incident radiation, so that two lines far apart in the spectrum cannot be compared directly. In analysing several spectral lines distributed over the entire spectrum several characteristic curves may be needed. For the highest precision a characteristic curve must be determined for each individual photographic plate.

7.4. Qualitative analysis

The direct current arc provides a rapid and very sensitive method for qualitative inorganic analysis. Rare gases, permanent gases and halogens cannot be detected, but nearly all other elements emit strong spectral lines in the arc, and several can be detected at concentrations of 10^{-4} per cent in the solid sample (Table 7.2).

Electrically conductive substances, such as alloys, can be used directly as electrodes provided that sufficient material is available.

TABLE 7.2. Analysis lines and limits of detection for 1 mg sample in the direct current arc with graphite electrodes

Element	*Analysis line (wavelength in nm)*	*Limit of detection in ppm*
Ag	328·0	0·5
Al	394·4	1
B	249·8	1
Be	234·8	2
Cd	326·1	50
Co	340·5	5
Cu	327·3	0·1
Fe	302·0	10
Mg	285·2	0·01
Mn	279·4	2
Mo	313·3	5
Pb	283·3	10
Sn	284·0	5
Ti	337·3	5
V	318·4	10
W	294·4	100
Zn	334·5	100

To prevent excessive melting of the electrodes, the arc current is then usually limited to 2–4 A. Non-conductive materials are powdered and a few milligrams introduced into a graphite electrode. Solutions may also be analysed by putting a few droplets on a porous electrode and letting the solvent evaporate.

A disadvantage of graphite electrodes is the obliteration of the spectral region between 360 and 460 nm by strong molecular emission bands due to the CN produced in the reaction between evaporated carbon and nitrogen from the air. These bands may be suppressed completely by burning the arc in an atmosphere of argon or carbon dioxide, to which some pure oxygen is added to maintain steady burning of the electrodes.

Qualitative analysis is carried out by checking the sample spectrum for the presence of the characteristic lines of an element. The bottom spectrum of Fig. 7.3 shows a few spectral lines of the

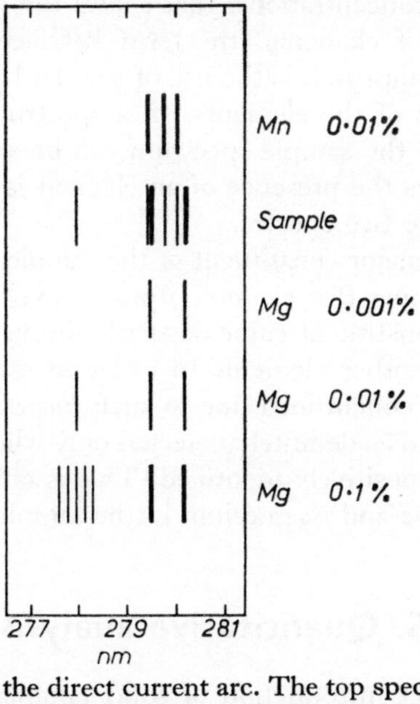

FIG. 7.3. Analysis with the direct current arc. The top spectrum shows the three persistent lines of manganese; the three spectra from the bottom upwards illustrate the decrease of the intensity of magnesium lines with decreasing concentration of magnesium. Spectral lines of manganese as well as magnesium are readily observed in the sample spectrum.

element magnesium. As the concentration of this element in the sample decreases, the number of lines decreases, until only the two or three strongest remain visible; these are the lines with a large transition probability (§ 2.6) and a low excitation energy, and are usually resonance lines (§ 2.2). Such lines are called *persistent lines* (*raies ultimes*, *Letzte Linien*). Fig. 7.3 shows the three persistent lines of manganese at 280 nm (a triplet) as well as two strong lines of magnesium (the persistent line of magnesium lies at 285 nm). Obviously, a monochromator of high resolving power is required. By comparing the sample spectrum with these standards the presence of both manganese and magnesium in the sample is readily confirmed.

It is obviously impossible to record the spectra of each element separately for every analysis. One solution is to make a spectrum of commercially available R.U. powder, a mixture of oxides, in suitable concentrations, that shows the persistent lines of a large number of elements (the term R.U. comes from *raies ultimes*). Another approach is the use of standard plates with the characteristic lines of the elements. In a spectrum projector the standard plate and the sample spectrum can be viewed simultaneously. In both cases the presence of an element is rapidly verified by comparing the two spectra.

If the major constituent of the sample shows a large number of spectral lines (for example iron), it may happen that a weak line of this constituent coincides with and is mistaken for a persistent line of another element. In order to minimize the possibility of incorrect conclusions due to such *line interferences*, an element is considered as definitely detected only when two or three persistent lines are positively identified. This is obviously the case for both manganese and magnesium in the sample spectrum of Fig. 7.3.

7.5. Quantitative analysis with the arc

The direct introduction of solid samples into the electrode, the small amount of sample required, and the great sensitivity of the arc spectrum warrant a closer inspection of the applicability of the direct current arc to quantitative analysis. The spectra in Fig. 7.3 show that the 'blackness' of the magnesium lines increases with the concentration of that element in the sample. From a visual comparison of the magnesium lines of the standards with those of the sample spectrum it can be concluded that the magnesium content of the sample is definitely larger than 0·001 per cent, but slightly less than 0·01 per cent. The manganese content is also less than 0·01 per cent. This example illustrates how arc spectra can be used to obtain a rapid, *semi-quantitative* estimate of the concentration of an element in the sample.

For a truly quantitative determination the limits must obviously be drawn much closer. Unfortunately, quantitative analysis with the direct current arc is subject to appreciable random and systematic errors, and several precautions must be taken to reach a precision of 10 per cent.

The major problem is the volatilization of the solid sample. The conversion of the concentration c of an element in the electrode to the concentration n of that element in the discharge region may be denoted formally by

$$n = Kc \tag{7.1}$$

(Compare Equation 6.1.)

However, volatilization is a complicated process that depends upon the arc current, upon the geometry of the electrodes and upon the chemical and physical properties of the sample. Complex chemical reactions occur in the hot graphite electrode. Atoms may diffuse through the porous electrode walls or be ejected sideways. The erratic volatilization is characterized by sudden eruptions interspersed with smooth periods. Volatile elements (Zn, Pb) evaporate more rapidly than less volatile elements (W, Zr), a phenomenon known as *fractional distillation*.* Consequently, unless precautions are taken, the proportionality factor K in Equation 7.1 is highly irreproducible between successive exposures.

Various techniques are applied to control the volatilization and to improve the precision and accuracy of the determination. Chemical and physical differences between synthetic standards

* In the analysis of high-purity uranium compounds used in nuclear power plants, fractional distillation is deliberately enhanced by using special electrodes and by adding so-called *carriers* (Ga_2O_3, AgF), which enhance the volatilization of the impurities but suppress the extremely rich spectrum of uranium.

(used to prepare the analytical curve) and natural products are removed by fusing all samples with lithium metaborate ($LiBO_2$) prior to the analysis. The fused product is powdered, homogenized, and mixed with excess graphite powder which promotes a smooth evaporation from the thin electrode. Alternatively, the sample may be pressed into a firm tablet with excess graphite powder.

Whereas dissociation of the sample molecules is virtually complete in the arc plasma, several elements are significantly ionized, so that the instantaneous emission intensity of an *atomic* spectral line is given by

$$I = \frac{Agh\nu b(1-\alpha)n}{B} e^{-E/kT} \qquad (7.2)$$

In this equation, which should be compared with Equation 6.5, A (the transition probability), g, h, ν, B, k (Boltzmann's constant), and E (the excitation energy) are constant for a particular spectral line; b is the (effective) diameter of the arc discharge, α is the degree of ionization and T is the arc temperature, which according to Table 7.1 is of the order of 5000–6000°K.*

It is readily calculated from Equation 7.2 that, all other conditions being equal, the intensity of a spectral line with an excitation energy of 4 eV (a resonance line at 310 nm) approximately doubles when the arc temperature is raised from 5600 to 6100°K. Now, as we saw in § 7.2, the arc temperature is determined primarily by the major constituent of the sample charged into the electrode cavity. Hence, similar concentrations of an element would yield a much higher intensity in an aluminium matrix (T = 6100°K) than in a lithium matrix (T = 5600°K). Consequently, it is imperative to control the arc temperature by the addition to all samples of a *spectrochemical buffer* that supplies the electrons for the discharge and minimizes the influence of the sample. For example, when the fusion technique discussed above is applied, the added lithium serves to fix the temperature at a value of 5600°K. This

* The reader may wonder how the high temperatures of flames and arcs are measured. One popular method is based upon Equation 7.2. The intensity ratio of two atomic spectral lines of one and the same element is given by $I_1/I_2 = (A_1g_1\nu_1/A_2g_2\nu_2)\exp[(E_2-E_1)/kT]$. When two lines are selected with known values for E, ν, A and g, the temperature T can be calculated directly from the measured intensity ratio.

tends to make both the degree of ionization α and the exponential factor in Equation 7.2 constant. The moderate temperature of the lithium arc suppresses the background intensity of molecular bands while maintaining a sufficient intensity of most atomic spectral lines.

One final technique often used to improve the precision of quantitative analysis is the application of the *internal standard principle*. The intensity of a spectral line is not considered alone, but instead is always compared to the intensity of a line of a reference element. This internal standard is added in equal, small amounts to all samples and standards. According to Equations 7.1 and 7.2 the intensity ratio of the two lines can then be written as

$$\frac{I_{\text{an}}}{I_{\text{i.s.}}} = \frac{K'_{\text{an}}(1-\alpha_{\text{an}})c_{\text{an}}}{K'_{\text{i.s.}}(1-\alpha_{\text{i.s.}})c_{\text{i.s.}}} \exp[(E_{\text{i.s.}}-E_{\text{an}})/kT] \qquad (7.3)$$

where the subscript 'an' denotes the analysis line of the element to be determined and 'i.s.' refers to the line of the internal standard.

The internal standard must be chosen in such a way that the intensity ratio expressed by Equation 7.3 is less sensitive to random fluctuations of the arc parameters than the single intensity given by Equation 7.2. To this end, the excitation energies, the ionization energies and the volatilization properties of the analysis line and the internal standard line should be matched. Also, the wavelengths of the two lines must be close enough to allow the use of the same characteristic curve for both lines (§ 7.3). Obviously, the internal standard element must not be present in the original sample. Therefore, selection of an internal standard is by no means a simple matter and presently there are, perhaps, just as many arc analyses carried out without an internal standard as with an internal standard. Internal standards are widely used in spark analyses (§ 7.7) and occasionally in other fields of spectrometry, such as flame spectrometry and X-ray spectrometry (where it is known as *ratio recording*).

As has been remarked in § 7.3 the quantity obtained from the photographic density S of a spectral line by means of the characteristic curve is the logarithm of the exposure:

$$\log H = \log \int I\,\mathrm{d}t \simeq \log It_e = \log I + \log t_e$$

where t_e is the exposure time. Accordingly, the analytical curve in arc spectrometry is usually expressed in logarithmic form as

$$\log H = p \log c_{an} + q \tag{7.4}$$

where p and q are constants. Theoretically the slope p of the analytical curve should be equal to unity, but in practice it is often smaller. One reason for this is self-absorption (§ 6.3).

With the provisions discussed above, a precision of 10 per cent can be reached in quantitative analyses with the direct current arc. The absence of systematic errors may be verified by checking the analytical procedure with reference samples of known composition. In view of the limited precision quantitative applications of the direct current arc are mainly restricted to those cases in which the amount of sample available (< 1 mg), the attainable detection limits (Table 7.2) or the nature of the sample (geological substances) make its use profitable.

7.6. The spark discharge

Fig. 7.4 shows the simplified and mutually equivalent electrical circuits of the *low-voltage* and the *high-voltage spark discharge.* Let us first consider the independent high-voltage spark, in which case the circuit on the right is disconnected.

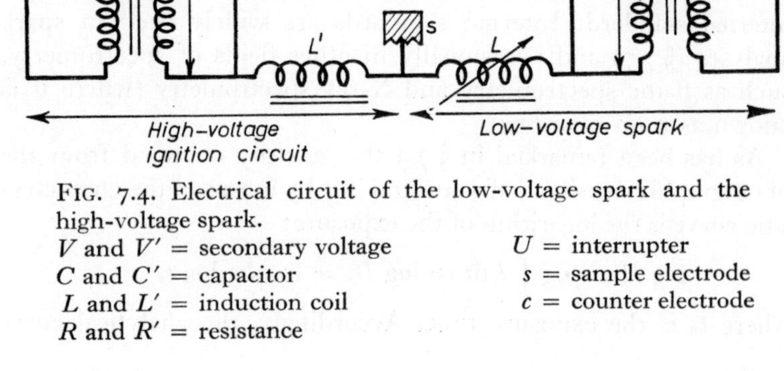

FIG. 7.4. Electrical circuit of the low-voltage spark and the high-voltage spark.

V and V' = secondary voltage
C and C' = capacitor
L and L' = induction coil
R and R' = resistance
U = interrupter
s = sample electrode
c = counter electrode

The secondary winding of the transformer yields an alternating voltage with a peak value of 20000 V that charges the capacitor C'. When the circuit is closed by the interrupter U, this potential difference also stands across the electrodes. However, the air gap of a few millimetres cannot withstand such a high voltage and, as a result, a pulse of electrons is released by the cathode. Because of the high potential difference, the electrons rapidly gain a high kinetic energy. Through collisions energy is transferred to the air molecules between the electrodes, and multiply charged positive ions are formed. These ions move rapidly towards the cathode where the vigorous impact of the heavy ions breaks sample away from the cathode. All this occurs within less than a microsecond.

With the release of the electrons the capacitor C' is discharged, but it is immediately charged again, this time with reversed polarity. When next the capacitor discharges again the process is repeated in the opposite direction. Clearly, the electrons and also the tiny spot of sample are now released by the other electrode. In this way, the electrons move back and forth and the electrical current during a single high-voltage spark discharge shows the well-known pattern of a damped oscillation presented in Fig. 7.5a.

Initially, the peak values of the current are very high and the temperature can reach values of 40000°K. Accordingly, the sample atoms are converted into multiply charged ions. Gradually, and in dependence on the value of the decay constant $2L/R$, the peak current and the temperature decrease, singly charged ions and atoms are formed, until finally the intensity of all spectral lines becomes zero and the discharge dies out.

However, the total duration of a single high-voltage spark is only some 100 microseconds and within that period the secondary voltage V' has hardly changed (note that a half cycle of the 50 Hz mains lasts 10000 μs!). As a result of the ions lingering between the electrodes, a new spark will immediately be initiated, but it will not be similar to the first one. If the interrupter in Fig. 7.4 is permanently closed, a series of erratic sparks is produced and the intensity of the spectral lines is hardly reproducible. To improve the precision of the analysis, the spark discharge is *controlled* by closing the interrupter only for a very short period and deliberately terminating the spark after 100 μs. The ionization of the gas cloud between the electrodes is allowed to disappear completely before a new, independent spark is generated by closing the interrupter again. In modern instruments the rotating interrupter shown in Fig. 7.4 is replaced by an electronic interrupter that permits a precise control of the discharge.

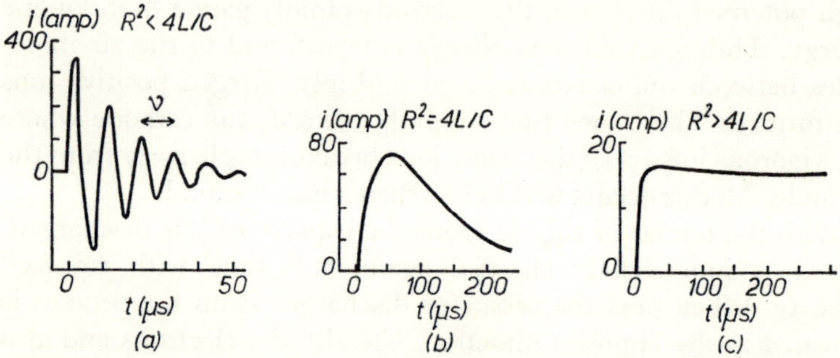

High-voltage spark	*Low-voltage spark*	*Low-voltage spark*
$V' = 20\,000$ V	$V = 1000$ V	$V = 1000$ V
$R' = 5\ \Omega$	$R = 10\ \Omega$	$R = 60\ \Omega$
$L' = 50\ \mu$H	$L = 300\ \mu$H	$L = 300\ \mu$H
$C' = 0{\cdot}02\ \mu$F	$C = 10\ \mu$F	$C = 60\ \mu$F
$O = 1$ MHz		

FIG. 7.5. Current variation during a single spark. (a) Damped oscillation; the values given for the electrical parameters refer to the analytically used high-voltage spark. (b) Critical damping with electrical parameters for a low-voltage spark. (c) Overdamped low-voltage spark.
N.B. Note the difference of the scales in the three figures.

In practice, between one and ten sparks are generated in each half cycle of the mains voltage. This amounts to several hundred sparks per second and, with an exposure time of thirty seconds, the spectral line intensity is the integral over several thousands of mutually independent sparks. It will be clear that this integral is hardly affected by an occasional spark of double or even tenfold intensity. Therefore, quantitative spark analyses tend to be much more reproducible than arc analyses.

Let us now turn to the low-voltage spark, in which case the entire circuit of Fig. 7.4 is used. Again, the capacitor C is charged by the secondary voltage V, but this time the peak value is only some 1000 V, which is insufficient to break down the electrode gap. Just as for an arc discharge a little initial ionization of the air is needed, and this is provided by the ***trigger spark*** of the high-voltage circuit. It is now essential, however, that the capacitance C' be very small so that the primary ignition spark does not contribute to the spectra emitted between the electrodes.

If for the electrical parameters of the low-voltage spark we had $R^2 < 4L/C$, as for the independent high-voltage spark, then the result would again be a damped oscillatory electrical current. Generally, however, the electrical parameters are chosen in such a way that $R^2 \simeq 4L/C$. Now, the spark discharge is *critically damped*, and Fig. 7.5b shows that instead of an oscillatory pattern the spark consists of a single current pulse. Consequently, the pulse of electrons moves only in one direction through the discharge zone, and the sample is broken away from only one electrode. By closing the interrupter and thereby triggering the *unidirectional spark* discharge only in the appropriate half cycle of the main voltage, it is possible to release the electrons and material only from the sample electrode. This arrangement is usually preferred in metallurgical analysis, where the cathode consists of the alloy to be analysed whereas the counter electrode, a tungsten or silver rod, does not contribute to the spectrum.

As is clear from Fig. 7.5b the current during a critically damped spark discharge is fairly low and, therefore, the plasma temperature is only moderately high. Critical damping of the spark discharge promotes the intensity of monovalent ionic lines and atomic lines in the far ultra-violet region of the spectrum, which is especially important for the analysis of ferrous alloys (§ 7.7). Fig. 7.5b also shows that the critically damped current pulse dies out fairly rapidly, and, just as for the high-voltage spark, the discharge is deliberately ended by opening the interrupter U.

If, finally, the electrical parameters are chosen so that $R^2 > 4\,L/C$, the spark discharge is *overdamped*. Fig. 7.5c shows that the current is even further reduced and remains practically constant until the interrupter ends the discharge. The overdamped low-voltage spark thus consists of a rapid repetition of very brief arc discharges, and in fact this type of discharge is also known as a *condensed arc*. In this way strong heating and melting of the electrodes is prevented, but the spectrum obtained closely resembles the spectrum of an arc discharge, although at lower intensity.

7.7. Spark analysis

For quantitative analysis the spark discharge is often preferred over the arc discharge in view of its much greater precision, which results from the integration over a large number of independent sparks, as explained in the preceding section. On the other hand, the spark impact ejects only very little material from the electrode (< 1 μg), so that the spark discharge should not be considered for trace analysis or for qualitative analysis.

The high-voltage spark is mainly applied to solution analysis, and the usual concentration range is 0·1–10 per cent as referred to the original solid sample. In the *rotating disk* technique a graphite disk is rotated slowly through the solution while it is being sparked against another graphite electrode. In the *copper spark* technique a large droplet is brought onto a copper electrode, the solvent is evaporated and the residue is sparked against another copper electrode; this is an example of microanalysis as the total amount of sample is only 100 μg. A conventional spectrograph is used, and the precision of the analysis (utilizing the internal standard principle) is 1–3 per cent.

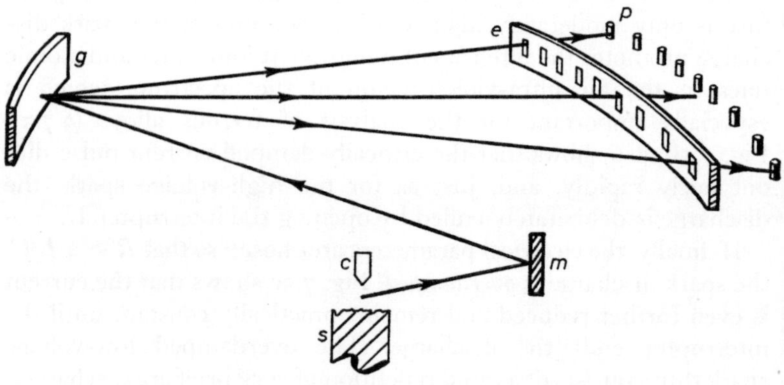

FIG. 7.6. Schematic drawing of a direct-reading spark spectrometer.

c = counter electrode
s = sample electrode
m = plane mirror
g = concave grating
e = exit slits
p = photomultipliers

In the important application of spark analysis in the metallurgical (notably steel) industry a combination of a unidirectional low-voltage spark and photo-electric detection is used with an instrument that is shown schematically in Fig. 7.6. A large, flattened portion of the alloy is used as the sample electrode, and a pointed tungsten or silver rod serves as the counter electrode. For steel analyses a vacuum monochromator is used, because the important elements, sulphur, phosphorus and carbon, are determined by means of spectral lines with wavelengths smaller than 200 nm (§ 4.4). The air-tight electrode chamber is flushed with argon.

A concave reflection grating in fixed position focuses the radiation and provides a large uniform dispersion over the entire spectrum. In the image plane of the concave grating a number of exit slits is arranged in such a way that each slit receives radiation of a single, preselected spectral line of an element of interest. A photomultiplier behind the exit slit converts the light pulse from the spark discharge to a current pulse that charges a capacitor. After exposure for about 30 seconds the electrical charges accumulated on the capacitors are measured; these are proportional to the time-integrated intensities.

In the wide spectrum of the high-dispersion monochromator up to thirty exit slits can be accommodated. One of these is adjusted to a spectral line of the major constituent of the alloy and controls the exposure time, the exposure being ended automatically when the charge on the corresponding capacitor reaches a predetermined value. A few other spectral lines of the major constituent serve as internal standard lines (§ 7.5). The outputs of the capacitors may be fed into a computer that derives the concentrations of the elements from the collected electrical charges, according to a prescribed computer program. In this way, the complete analysis of a sample can be made available within a few minutes. Of course, the adjustment and the programming of such expensive instruments take a long time.

The entire equipment is very suitable for the rapid and precise determination of fixed sample constituents in concentrations exceeding 0·01 per cent. With the application of the internal standard principle, the random error can be reduced to 0·3 per cent relative. On the other hand there are several potential sources of systematic errors.

If the sample is taken from a huge amount of material, as is the case in a steel foundry, care must be taken to ensure that it is representative of the bulk and that its composition does not change during transport from the foundry to the spectrometer. It should further be noted that the spark impact is restricted to the surface of the electrode. Ultimately, the quality of several tons of steel is judged on the basis of an analysis of a mere milligram!

The strong local heating due to spark impacts may alter the surface of the electrode and cause variations of the spectral intensity during sparking. These 'sparking-off' effects depend upon the sample, upon the element in question, and upon the values of the electrical parameters. Accordingly, in developing a procedure for routine analyses the electrical quantities (inductance, capacitance and resistance) must be optimized for the sample in concern.

Several precautions are taken to avoid or minimize systematic errors in the determination. In order to obtain a reproducible electrode surface the solidified sample is cut and polished. The capricious sparking-off effects during the first few seconds of the sparking process are avoided by starting the exposure after a 'pre-spark time' of the order of 10 seconds. The complete analytical procedure is extensively tested by means of reference samples of known composition, some of which are commercially available. The proper functioning of the instrument is checked periodically by analysing a standard sample; this also permits compensation for drift (§ 1.6). Finally, samples can occasionally be analysed by other, slower methods and the results can be mutually compared.

7.8. Some other sources of excitation

The ideal source of excitation for atomic spectrometry would combine the stability and freedom from interferences of the flame with the precision of the spark and the versatility of the arc discharge. Over the past ten years several attempts have been made to develop such a source and although none of these attempts has been completely successful as yet, some of the proposals will be briefly indicated.

Chemical flames with high temperatures are known, but are hazardous in normal operation (C_2N_2–O_2 or H_2–F_2); moreover, the temperature of the pure flame (4500°K) is decreased to about 3500°K when aqueous solutions are aspirated into it.

Stabilization of an arc discharge by blowing a stream of argon along the electrodes has been only moderately successful. Although the reproducibility of the analysis is somewhat improved, the rapid fluctuations are still much too large to allow instantaneous photoelectric recording.

A substantial improvement is obtained if the sample is introduced in liquid form rather than in solid form. Of course, the necessity of dissolving the sample prior to analysis is a serious limitation (§ 6.6). In the *plasma jet* the sample solution is aspirated through a hole in the lower electrode of an arc discharge. In some complicated arrangements the instantaneous fluctuations in intensity can be reduced to about 5 per cent, which is still larger than those observed in flame emission spectrometry.

The *induction-coupled plasma* appears to be more promising. In this device, the solution is aspirated by argon gas into a silica tube with a diameter of a few centimetres. A coil around the tube provides a strong, alternating electromagnetic field with a frequency of the order of 30 Mc/s. The energy of the radiofrequency waves is transferred to the gas, the solvent evaporates, and the atoms are excited (this is comparable to the microwave excitation in the source of radiation mentioned in § 6.5). Preliminary measurements indicate a high temperature (6000°K), a stable emission intensity, and low limits of detection.

Bibliography

1. AHRENS, L. H., and TAYLOR, S. R., *Spectrochemical Analysis*, Addison-Wesley, Reading, Mass., 2nd ed., 1961.
2. BOUMANS, P. W. J. M., *Theory of Spectrochemical Excitation*, Adam Hilger Ltd, London, 1966.
3. HARVEY, C. E., *Spectrochemical Procedures*, Applied Research Laboratories, Glendale, Calif., 1950.
4. HELLER, B., *La Spectrochimie Photo-électrique*, Soc. Fr. d'instruments de contrôle et d'analyses, Le Mesnil, 1968.
5. AMERICAN SOCIETY FOR TESTING MATERIALS, *Methods for emission Spectrochemical Analysis*, 5th ed., Philadelphia, 1968.

Problems

1. The electrical circuit of a direct current arc contains a 250 V source. The voltage drop across the electrodes is 50 V. If the current decreases by one ampere, the voltage drop is observed to increase to 70 V. What is the minimum value of the series resistance required to obtain a stable discharge?

2. In the construction of a characteristic curve a mercury lamp of constant intensity is used. The exposure time is equal for all spectra, and the exposure is varied by placing filters of known transmittance between the lamp and the spectrograph. In each spectrum the same mercury line is measured with a microphotometer. The following results are obtained:

Filter no.	*Transmittance of filter* (%)	*Transmittance of spectral line* (%)
—	100	6·3
1	60	10·0
2	40	19·0
3	20	50·0
4	15	66·0
5	10	79·5

Determine the characteristic curve. What is the gamma of this emulsion?

3. Continuing the construction of the characteristic curve from problem 2, spectra are now recorded of five standard tin samples, to which antimony is added as an internal standard. The microphotometer now yields the following data:

Concentration Sn (%)	*Transmittance Sn line* (%)	*Transmittance Sb line* (%)
0·1	7·2	32
0·05	11·5	26·2
0·02	31·6	25
0·01	67·5	28·2
0·005	72·5	18·2

Calculate the densities of the lines and derive the corresponding logarithmic exposures from the characteristic curve. Finally, plot the analytical curves with and without the internal standard. Does the use of an internal standard improve the result?

4. The analytical curve of problem 3 is tested with a standard sample. To this end, an ore containing 0·032 per cent Sn is analysed four times, according to the procedure given in problem 3. The results are:

Analysis	*Transmittance Sn line* (%)	*Transmittance Sb line* (%)
1	23	23
2	14·8	21
3	25·6	30
4	17·8	27

Calculate the experimental average concentration and its standard deviation. Does the determination show a systematic error?

5. Assume that a solid sample can be dissolved in water to a concentration of 1 per cent by weight. Consider that a minimum volume of 5 ml of sample solution is required in flame spectrometry, whereas 1 mg of solid sample is used in arc spectrometry. Use the data in Tables 6.2 and 6.3, with manganese as an example, to compare the *relative* limits of detection (ppm Mn in the solid sample) and the *absolute* limits of detection (in ng Mn) reached by both methods of analysis. Also comment on the absolute amount of sample needed in both cases.

6. Draw an electrical circuit for the direct current arc with high-voltage ignition. What interrupter would you use?

7. In arc spectrometry metallic electrodes melt away rapidly unless the current is limited to 2 A. Why does this phenomenon not occur in spark spectrometry?

CHAPTER EIGHT

X-RAY SPECTROMETRY

8.1. Continuum and characteristic X-ray spectra

Fig. 8.1 shows the principal components of a source for exciting X-rays. The *X-ray tube* is rarefied to a pressure of 10^{-7} atm. A thin wire with electrical leads sticking through the glass envelope is heated by a current of a few mA, whereupon the hot wire emits electrons. The electrons are accelerated under the influence of a large potential difference of, for example, 50000 V, and hit the

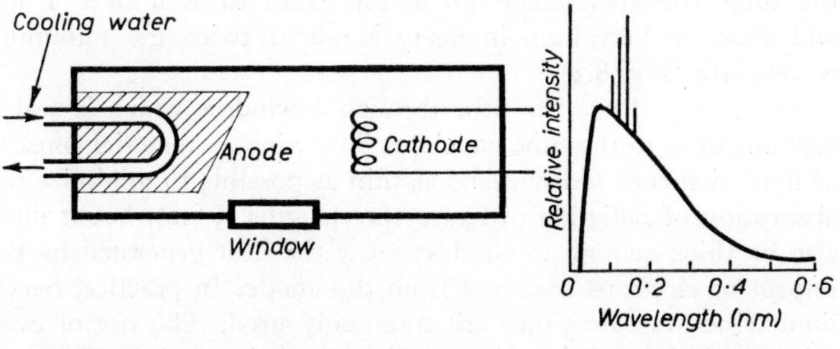

FIG. 8.1. Basic outline of an X-ray tube, and the intensity distribution of the spectrum emitted by a tube with a potential difference of 50 kV between the cathode and a tungsten anode.

target anode at a very high speed. The great majority of the electrons lose their kinetic energy by simply heating the target and, therefore, the anode must be cooled to prevent melting and volatilization of the target material.

About one per cent of the incident electrons are retarded by the atoms in the anode in such a way that their kinetic energy is partly converted to electromagnetic radiation. This radiation was first observed in 1895 by Röntgen, who called it X-radiation to emphasize its mysterious properties, notably its high penetrating power. The electromagnetic nature of the radiation was soon established, and in German (and Dutch) literature this part of the electromagnetic spectrum is called *Röntgenstrahlung*, after its discoverer. The name X-rays has persisted, however, in English and French literature.

The minimum wavelength of the radiation emitted by the tube is readily calculated from the potential difference across the tube. If the tube voltage is 50000 V, the kinetic energy of an electron at the moment of impact is 50 keV. When this energy is converted completely to electromagnetic radiation, the maximum photon energy is also 50 keV and the corresponding photon wavelength is 0·0248 nm or 0·248 Å (Appendix II). As, however, the electrons can release any arbitrary fraction of their kinetic energy as photons, the X-ray tube emits a broad *continuum spectrum* which ranges from a minimum wavelength determined by the tube voltage (0·0248 nm in this case) to well over 1 nm and shows a maximum intensity at about twice the minimum wavelength (Fig. 8.1).

The X-rays leave the tube through a window which is sealed vacuum-tight to the tube envelope. The window material consists of light elements and must be as thin as possible to minimize the absorption of radiation of larger wavelengths (§ 8.2). Yet it must also be thick enough to conduct away the heat generated by the impact of electrons scattered from the anode. In practice, beryllium windows of 0·3 mm are commonly used. The use of even thinner windows of organic material (consisting mainly of carbon and hydrogen) has been proposed for the determination of elements with atomic number 13 (Al) or less. Another possibility is the use of windowless tubes, but the attendant requirement of keeping the entire spectrometer at a pressure of 10^{-7} atm. forms a considerable practical objection to their application.

The relative distribution of the continuum ('white') X-ray spectrum is independent of the target material, but the overall intensity increases rapidly with the tube power (several kW) and with the atomic number Z of the target material. A tungsten anode ($Z = 74$) is frequently employed, because it yields an intense continuum spectrum. In addition to the continuum spectrum several sharp lines are observed in the emission spectrum that are characteristic for the target material used. For example, with tungsten, *characteristic lines* are found at 0·1476, 0·1282, 0·1098 and 0·1678 nm. The relative contribution of the characteristic lines to the tube spectrum increases with decreasing atomic number of the anode material. Thus, a copper anode ($Z = 29$) emits two strong characteristic lines at 0·15 nm, superimposed upon a relatively weak continuum background.

In order to explain the origin of the characteristic lines, we recall the model of the atom discussed in §§ 2.1 and 2.2. In heavier atoms the electrons are distributed over several shells, designated as the K-shell, the L-shell, the M-shell etc., according to the value of the principal quantum number of the electron ($n = 1, 2, 3$ etc.). With increasing principal quantum number the average distance of the electron from the atomic nucleus (§ 9.1) increases, and as a result the electrical attraction between the nucleus and the electron decreases. Comparatively little energy is required to remove the outermost valence electrons (ionization) or to raise them to higher energy levels (2–20 eV).

The binding energy of the K-electrons in an atom increases with the positive charge upon the atomic nucleus (for hydrogen the K-electron is also the valence electron), and varies from 1 keV for sodium to 75 keV for lead (nuclear charge of 82 elementary units). Fig. 8.2a shows the binding energies of the electrons in the inner shells of the molybdenum atom (atomic number 42). Incidentally, note that the transitions of the valence electron of molybdenum (Plate 1a) are restricted to a thin layer of 0·01 keV just below the ionization limit! An energy of 20 keV is required to remove an electron from the K-shell of molybdenum. This may occur in the molybdenum anode of an X-ray tube, where the electrons arrive with an energy of 50 keV. However, this process can also take place when molybdenum is irradiated with X-rays of wavelength smaller than 0·062 nm (photon energy exceeding 20 keV).

The molybdenum atom is now ionized from the K-shell and is raised to an unstable level of extremely high energy. The vacant position in the K-shell will quickly be occupied by an electron

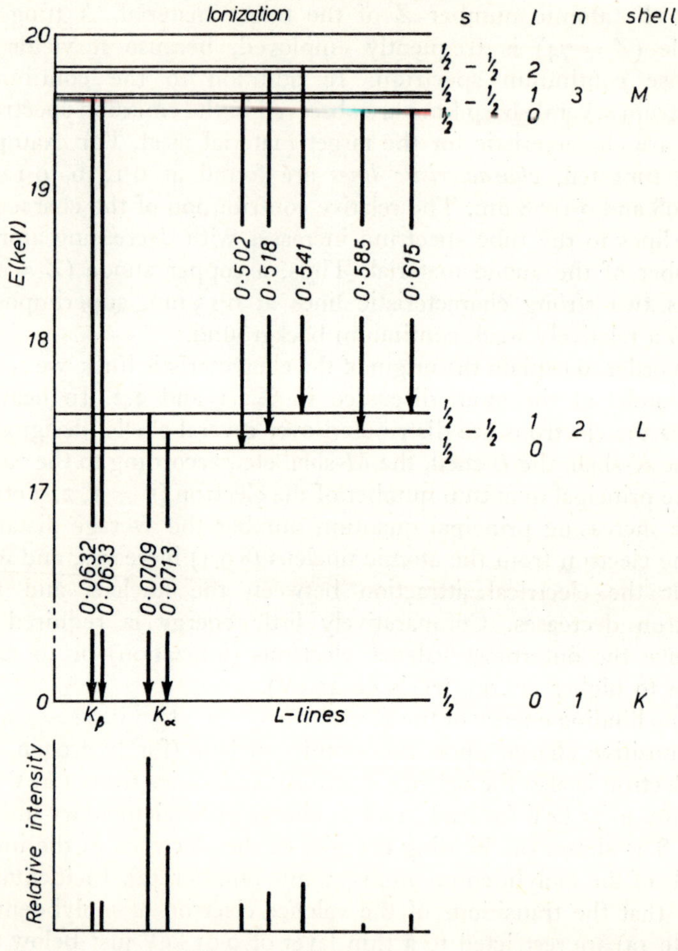

FIG. 8.2A. Level diagram of molybdenum with the wavelength of characteristic *K*-lines and *L*-lines in nm. The relative intensity of the transitions is also indicated.

from the *L*-shell or the *M*-shell, and the energy generated in this process may be lost in emitting a photon of electromagnetic radiation. This is called *K*-radiation. In exactly the same way the transition of an electron from the *M*-shell (or the *N*-shell) to a vacancy in the *L*-shell leads to the emission of *L*-radiation.

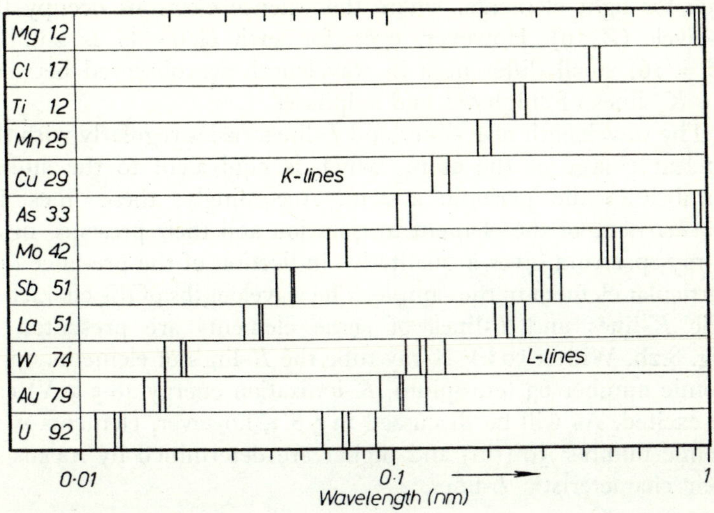

FIG. 8.2b. Wavelength of characteristic X-ray lines of some elements.

The *K*-shell consists of a single orbit ($n = 1$, $l = 0$), but the *L*-shell contains three different orbits, distinguished by the values of the orbital quantum number *l*, and the spin quantum number *s* (§ 2.2). The selection rule for a spectral transition is that the orbital quantum number must change by one unit ($\Delta l = \pm 1$). With this rule the various transitions of the molybdenum atom are readily derived (Fig. 8.2a). The K_α-lines, resulting from transitions between two *L*-levels and the *K*-level, show the greatest intensity. The photon energy of K_β-lines (*M*-levels to *K*-level- is slightly larger, and hence their wavelength is smaller; the intensity is five times smaller. The *L*-lines are much weaker than the *K*-lines.

For heavy elements such as molybdenum, the binding energy of the *K* and *L*-electrons, and hence the position of the *K* and *L*-levels, is independent of the distribution of the valence electrons. Accordingly, for these elements ($Z > 20$) the wavelength of the K_α-lines is independent of the physical or chemical environment (liquid or solid, pure or in compound form). Naturally, this is not true for light elements, where the valence electrons occupy the *L*-levels ($Z < 9$). However, even for such elements as sulphur ($Z = 16$) small differences in wavelength are observed between the K_α-lines of sulphates and sulphides.

The wavelength of *K*-lines and *L*-lines varies regularly with the nuclear charge of the atom, which is equivalent to the atomic number in the periodic system. Accordingly, these lines are *characteristic* of the element in question and their presence in an X-ray spectrum gives a qualitative indication of the presence of a particular element in the sample. The wavelengths of the characteristic *K*-lines and *L*-lines of some elements are presented in Fig. 8.2b. With a 50 kV X-ray tube the *K*-lines of elements up to atomic number 63 (europium, *K*-ionization energy 48·5 keV) can be excited. As will be discussed in § 8.3, however, elements with atomic number 50 (tin) and higher are determined by means of their characteristic *L*-lines.

8.2. Absorption and fluorescence of X-rays

For analytical purposes the excitation of characteristic X-rays by electron bombardment (as discussed in the preceding section) is less attractive, because simultaneously an appreciable amount of continuum background radiation is generated in this process. Upon irradiation with primary X-rays, however, only the characteristic lines of the sample elements are produced, and this method of analysis is appropriately called *X-ray fluorescence spectrometry*. Just as in molecular and atomic fluorescence of visible and ultraviolet radiation (§§ 4.8 and 6.5), the excitation of fluorescence radiation is preceded by the absorption of primary radiation. However, before we discuss the absorption of X-rays in more detail, let us reconsider the physical processes leading to the emission of fluorescence radiation.

The primary process is the ejection of an inner electron. The kinetic energy of this electron is equal to the difference between the photon energy of incident, monochromatic, X-rays and the electron binding energy. Thus the kinetic energies of the ejected electrons are characteristic of the elements in the sample, and their number is proportional to the concentration of an element in the sample. The question may therefore be asked, why not study the energy spectrum of the emitted electrons? However, the instrumental problems in measuring low-energy electrons are quite considerable, and one important restriction is that the electrons are only ejected from a very thin sample layer of a few hundred ångströms. Nevertheless, the possibilities of electron spectroscopy for chemical analysis (ESCA) are currently being seriously investigated. Promising results have been obtained in the study of light elements (compare § 8.6) where the energy spectrum of the ejected electrons may reveal the influence of chemical bonding of carbon, oxygen and nitrogen atoms, just as NMR spectrometry does for hydrogen atoms (Chapter 10).

After the ejection of an inner electron the ionized atom may use its excess energy to emit a characteristic X-photon, but it may also use this energy to release one or more valence electrons (the so-called *Auger* electrons). The relative probability of these two processes determines the *fluorescence yield factor* (§ 4.8), which is high for heavy atoms but decreases rapidly to values as low as 0·01 for light elements. Finally, on its way out of the sample the fluorescent X-ray will again be attenuated by secondary absorption. The absorption of X-rays thus plays a major role in quantitative X-ray spectrometry.

The absorption of X-rays can be described by an expression equivalent to the Beer–Lambert law for the absorption of optical radiation (§ 4.5):

$$I = I_0\, e^{-\mu^* b} = I_0\, e^{-\mu\rho b} \tag{8.1}$$

In this equation I_0 is the intensity of the incident (monochromatic) radiation and I is the intensity after a distance b has been traversed in the sample; μ^* is the *linear absorption coefficient* expressed in m^{-1} or cm^{-1}. This linear absorption coefficient depends upon the density of the sample, and is conveniently replaced by the product $\mu\rho$, where ρ is the density of the substance and μ is called the *mass absorption coefficient* (expressed in cm^2/g or m^2/kg).

The mass absorption coefficient depends only upon the wavelength of the X-rays and upon the atomic number of the absorbing element. Fig. 8.3 shows that μ increases quite strongly with the atomic number. Therefore, the windows of X-ray tubes and detectors (§ 8.4) are made of light elements, to minimize the absorption of long-wavelength radiation. Fig. 8.3 further shows that the mass absorption coefficient generally increases steadily with the wavelength, but a few abrupt decreases are observed; these are known as *absorption edges*. The origin of the discontinuities is readily understood by considering as an example the element chromium. The ionization energy required to remove an electron from the *K*-shell of chromium is equal to 6 keV. Photon energies of 12·4 keV are clearly capable of producing *K*-ionization, and accordingly X-rays with a wavelength of 0·1 nm are absorbed to some extent by chromium samples. As the wavelength increases, the photon energy approaches the ionization energy and the X-rays are more efficient in producing *K*-shell ionization. Indeed, the mass absorption coefficient of chromium is seen to increase tenfold when the wavelength increases from 0·1 to 0·21 nm (photon energy 6 keV). The photon energy of radiation with larger wavelengths is insufficient to produce *K*-shell ionization, and so the mass absorption coefficient drops abruptly, giving rise to the *K-absorption edge.* However, this radiation is still capable of ejecting an electron from the *L*-shell, and the absorption coefficient does not become exactly zero. With further increase of the wave-

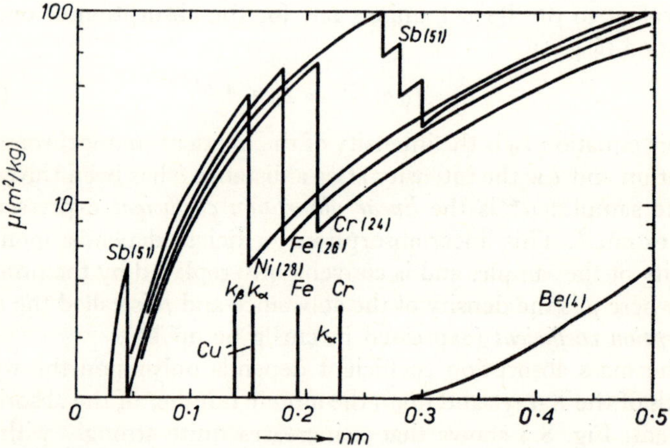

FIG. 8.3. The mass absorption coefficient of Be, Cr, Fe, Ni and Sb as a function of wavelength.

length the absorption coefficient increases again as L-ionization proceeds more efficiently. For antimony the K-absorption edge is found at 0·0406 nm, and the three L-absorption edges (corresponding to the three L-orbitals) are found at 0·264, 0·283, and 0·30 nm (Fig. 8.3).

The energy of an electron jumping from the L-shell to the K-shell is slightly less than the energy required for K-shell ionization (Fig. 8.2a). Consequently, the characteristic K-lines of an element are located just *beyond* the K-absorption edge. Each element shows comparatively little absorption for its own characteristic lines.

For multi-element samples (either mixtures or compounds) the mass absorption coefficient can be written as

$$\mu = \mu_1 W_1 + \mu_2 W_2 + \mu_3 W_3 + \ldots = \sum_i \mu_i W_i \qquad (8.2)$$

where μ_i and W_i are the mass absorption coefficient and the weight fraction of element i, respectively. For example, air contains 0·8 parts of nitrogen with a mass absorption coefficient of 5·5 m^2/kg at 0·3 nm, and 0·2 parts of oxygen with a mass absorption coefficient of 8·4 m^2/kg at this wavelength. The density of air is 1·3 kg/m^3. Combination of Equations 8.1 and 8.2 then shows that after a distance of 20 cm in air the transmittance for radiation of 0·3 nm is

$$\exp\left[-(0{\cdot}8 \times 5{\cdot}5 + 0{\cdot}2 \times 8{\cdot}4) \times 1{\cdot}3 \times 0{\cdot}2\right] = 0{\cdot}2$$

Therefore, the air present in an X-ray spectrometer would absorb 80 per cent of the 0·3 nm radiation. A similar calculation shows that for a wavelength of 0·6 nm the absorption is increased to 99·9 per cent! Consequently, for the observation of long wavelength radiation the air must be removed; in fact X-ray spectrometers operate at pressures of about 10^{-4} atm.

Because the density of solids and liquids is a thousand times larger than the density of air, the penetration of X-rays into condensed samples is very small. Depending upon the wavelength of the radiation and the presence of heavy elements in the sample, the *depth of penetration* varies from 5 to 500 μm.* Recall that the absorption attenuates both the primary incident beam and the beam

* These values explain the important application of X-rays in the measurement of the thickness of thin metal sheets.

of excited fluorescence radiation that must leave the sample (see problems 1 and 2). Consequently, and in marked contrast to optical fluorescence, the samples in X-ray fluorescence are nearly always *infinitely thick* and all of the incident X-rays are absorbed in a thin surface layer of the sample. This may be an advantage in surface studies. In general, however, the analyst must take great care to ensure that the composition of the surface of the sample agrees exactly with the bulk composition. Notably for powdered samples this is not easily achieved, and particle size effects are a notorious problem in X-ray spectrometry.

An approximate expression for the intensity of X-ray fluorescence radiation can be derived as follows. After a distance x in the sample the incident beam is attenuated according to Equation 8.1 to $I_x = I_0 \exp[-\mu\rho x]$. Over the next small distance $\mathrm{d}x$ the absorption by the element of interest is $\mathrm{d}I = \mu_i W_i \rho I_x$, of which a fraction Φ is re-emitted as fluorescence radiation. Because of secondary absorption the fluorescence beam is again attenuated, so that only a fraction $\mathrm{d}I_F = \exp[-\mu'\rho x]\,\mathrm{d}I$ escapes from the sample. (The symbol μ' is used to indicate that the wavelength of fluorescent radiation is generally different from the wavelength of the primary beam.) Consequently, the fluorescent intensity for an infinitely thick sample is found to be

$$I_F = \int_0^\infty \Phi \mu_i W_i I_0 \rho\, e^{-(\mu+\mu')\rho x}\,\mathrm{d}x = \frac{\Phi \mu_i W_i I_0}{\mu+\mu'} \qquad (8.3)$$

As expected, the fluorescent intensity is proportional to the fluorescent yield factor and to the incident intensity. For this reason high-power X-ray tubes are preferred, especially for trace analysis. Also, the fluorescent intensity is proportional to the weight fraction W_i, so that we expect linear analytical curves in X-ray spectrometry. However, closer inspection reveals that this can only be true in a first approximation, because according to Equation 8.2 the weight fraction also appears in the total mass absorption coefficients μ and μ' of the sample. Thus, when X-ray spectrometry is applied to the analysis of major constituents of a sample, the composition of the sample changes appreciably over the range covered by the analytical curve and, in turn, the mass absorption coefficients may change.

The determination of iron in chromium is a typical example. Fig. 8.3 show that the mass absorption coefficient for iron K_α-radiation (secondary absorption) is small for iron, but much larger for chromium. When the iron content of the standards is varied from 0 to 100 per cent, the mass absorption coefficient μ' steadily decreases and, hence, the fluorescent intensity increases more than linearly. Indeed, Fig. 8.4 shows that the analytical curve is concave. On the other hand, this effect does not occur in the

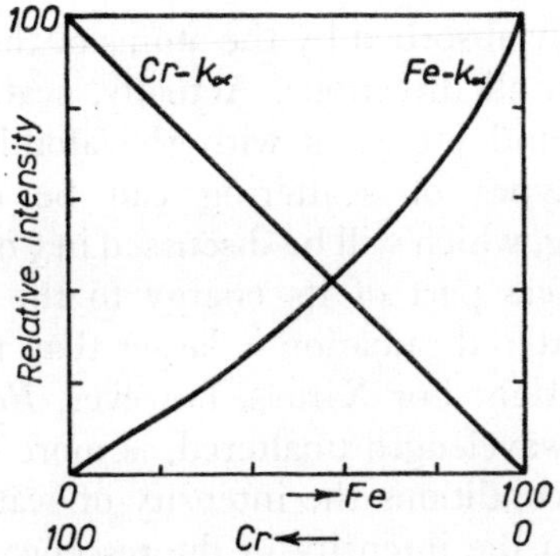

FIG. 8.4. Analytical curves for the determination of iron and chromium in mixtures of these two elements.

determination of chromium, because for *chromium* K_α-radiation the mass absorption coefficients of iron and chromium are approximately equal. Thus, μ' is constant in this case and the analytical curve for the determination of chromium in iron–chromium alloys is linear.

A more general problem is the variation of μ and μ' among different samples. Equal amounts of an element will yield a much higher intensity in a light aluminium matrix of small mass absorption coefficient than in a heavy tin matrix of large mass absorption coefficient. (Perceptive readers will note the similarity between this influence of the sample composition through the variable mass absorption coefficient, and its influence through the variable arc temperature discussed in § 7.5.) Two general procedures may be followed to minimize the possible systematic errors resulting from variable mass absorption coefficients. If samples of widely varying composition are to be analysed, all samples and standards are strongly diluted by dissolving in water or by fusion with borax to a glass. For routine analyses of a large number of similar samples it is more attractive to prepare the analytical curve by means of standards with a composition similar to that of the samples (§ 8.5).

8.3. Diffraction of X-rays by crystals

X-rays are not only absorbed by the atoms of the sample, but also partly scattered in all directions. Actually, scattering takes place at the electrons, and increases with the atomic number of the elements. Two types of scattering can be distinguished. In *Compton* scattering, which will be discussed in § 9.3 for γ-radiation, the photon transfers part of its energy to the electron and the wavelength of scattered radiation is larger than the wavelength of the incident radiation. For X-rays, however, *Rayleigh* scattering, which leaves the wavelength unaltered, is more important.

Under normal conditions the intensity of scattered radiation is much smaller than the intensity of fluorescence radiation excited in the sample. Scattering only contributes a weak continuum background to the X-ray spectrum. However, if the atoms are regularly spaced, as occurs in a crystal lattice, the scattered waves may enforce each other by constructive interference (§ 1.1) and the intensity of scattered radiation can be appreciably enhanced. This effect is used to study the structure of crystalline substances (§ 8.7). It is also used to separate the wavelengths of a polychromatic beam of X-rays.

A small part of a crystalline substance is shown, enormously enlarged, in Fig. 8.5. A parallel beam of X-rays strikes the surface of the crystal at an angle θ. (Note that the angle of incidence is taken to be equal to the angle between the incident ray and the crystal surface; this differs from the convention used in optical reflections, see Fig. 1.2.) The X-rays are scattered by the atoms in the crystal, but in most directions the scattered radiation will be quenched by destructive interference (§ 1.1). Positive enforcement occurs only for special wavelengths and in a particular direction.

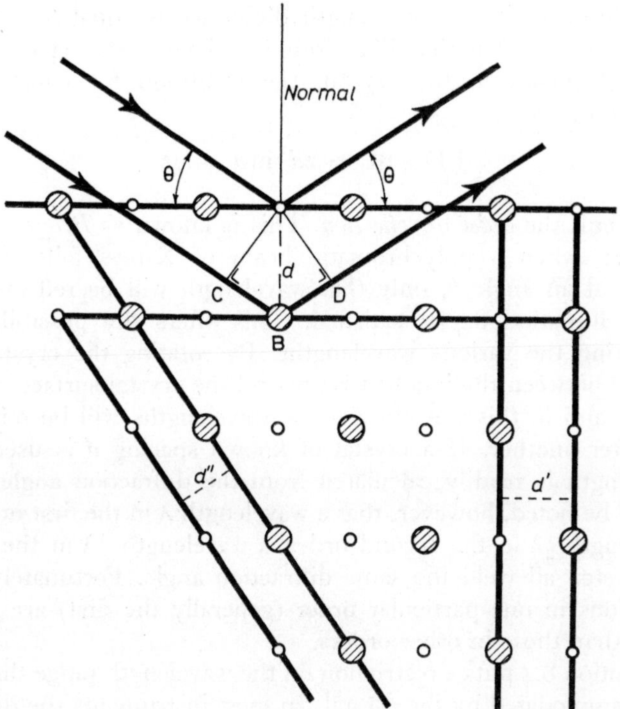

FIG. 8.5. Dispersion of X-rays by a crystal; d, d' and d'' are separations between successive atomic layers of three different sets of crystal planes.

Constructive interference of the waves scattered by the atoms *at the surface* is possible only if the incident ray, the *diffracted* ray and the normal to the surface all lie in one plane, and if the angle of 'reflection' is equal to the angle of incidence, θ. The X-rays also penetrate the crystal (a depth of penetration of 5 μm corresponds to at least 10000 atomic layers). Because the refractive index for X-rays is very close to unity, the incident beam is not refracted. According to Fig. 8.5 the path length difference of two rays diffracted by two successive atomic layers is equal to BD+BC. Just as for the diffraction of ultra-violet and infra-red radiation at gratings (§ 3.4), constructive interference of successive rays is possible only if this path length difference is equal to a whole number of wavelengths. Therefore, if d denotes the separation of adjacent planes in the crystal, the condition for constructive interference is

$$BD + BC = 2d \sin\theta = m\lambda \qquad (8.4)$$

where m is the *order of reflection*. This is known as *Bragg's Law*.

Thus, when a polychromatic beam of X-rays falls upon a crystal at an angle θ, only that wavelength will be reflected for which Equation 8.4 is satisfied. This offers the possibility of separating the various wavelengths. By rotating the crystal the angle θ between the incident beam and the crystal surface can be varied, and in this way the various wavelengths will be reflected one after another. If a crystal of known spacing d is used, the wavelength is readily calculated from the diffraction angle θ. It should be noted, however, that a wavelength λ in the first order, a wavelength $\frac{1}{2}\lambda$ in the second order, a wavelength $\frac{1}{3}\lambda$ in the third order, etc., all yield the same diffraction angle. Fortunately, the reflections in one particular order (generally the first) are much larger than those in other orders.

Equation 8.4 puts a restriction on the wavelength range that can be accommodated by the crystal. In most instruments the diffraction angle θ is limited to a range of 11° to 70° (see problem 3); accordingly (in the first order), the wavelength may vary between 0·38 d and 1·88 d. According to Table 8.1 crystals are available with spacings up to d = 1·3 nm, so that wavelengths up to 2·5 nm can be analysed (compare Fig. 8.2b). On the short

TABLE 8.1. Analysing crystals for X-ray fluorescence spectrometry

Crystal	d (*nm*)	λ_{min} (*nm*)	λ_{max} (*nm*)	*Elements covered* *K-lines*	*L-lines*
Topaz ($Al_2F_2SiO_4$)	0·136	0·052	0·256	V(23)–In(49)	Ce(58)–U(92)
LiF	0·201	0·076	0·378	K(19)–Zr(40)	In(49)–U(92)
Quartz	0·425	0·161	0·800	Si(14)–Ni(28)	Rb(37)–Lu(71)
$(NH_4)H_2PO_4$	0·532	0·202	1·00	Mg(12)–Mn(25)	As(33)–Gd(64)
Gypsum ($CaSO_4$)	0·760	0·29	1·43	Na(11)–Sc(21)	Cu(29)–I(53)
Potassium-hydrogen phthalate	1·32	0·50	2·48	F(9)–S(16)	V(23)–Mo(42)

wavelength side the situation is less favourable. The smallest value for d is 0·136 nm, whence the minimum wavelength is 0·052 nm. This corresponds to K_α-radiation of the element indium. In fact, this is the reason that elements heavier than indium are determined by their L-lines (§ 8.1).

8.4. Detection of X-rays

X-ray spectra may be recorded photographically. This is completely analogous to the photographic detection of visible and ultra-violet radiation discussed in § 7.3. The gelatin coating of the photographic emulsion must be extremely thin to prevent absorption of long-wavelength X-rays. However, the time-consuming processing of photographic films and the complicated relation between the incident intensity and the density of the photographic image are major objections, and consequently photographic detection is hardly ever used in analytical X-ray spectrometry. Instead, photo-electric detectors are employed, which convert the X-rays into an electrical current. These detectors are based upon the ionization produced by X-rays, and this property of X-rays can only be explained by the photon-character of electromagnetic radiation.

For a clear understanding of the following discussion it is essential to make a sharp distinction between the energy of *individual* photons and the *total* intensity of the beam. The *photon energy* is inversely proportional to the wavelength of the X-rays, and varies from 25 keV for a wavelength of 0·05 nm to 1 keV for a wavelength of 1·24 nm. The *total intensity* of the beam is determined by the rate of the incident photons, that is by the number of photons reaching the detector per second. Thus the detection of X-rays can best be described as the counting of individual photons. In the detector each photon gives rise to a *current pulse*. The total intensity of the X-ray beam is then represented by the *counting rate*, which indicates the number of electrical pulses measured per second. A similar treatment will be given in § 9.4 for the detection of γ-radiation.

In Fig. 8.6a an X-ray photon enters a tube filled with argon at a pressure of 0·1 atm. through a window of 15 μm thick mica or

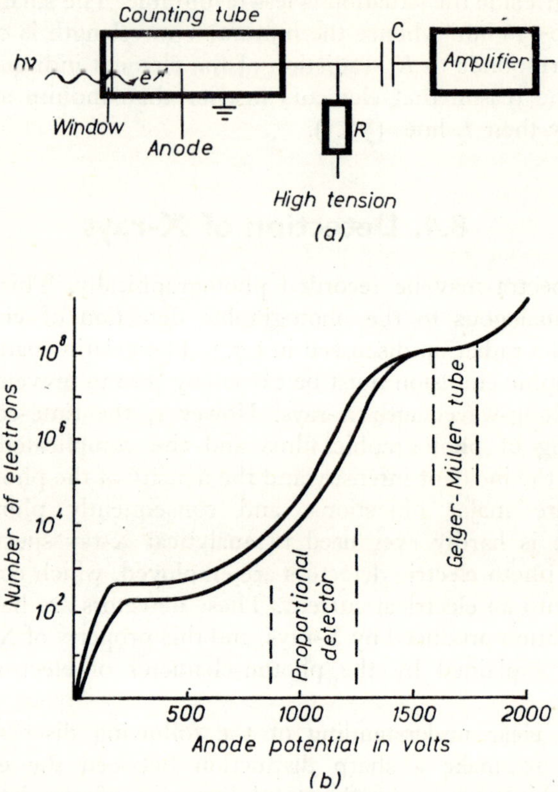

FIG. 8.6. (a) Diagram of a proportional detector. (b) Number of electrons reaching the anode of the detector as a function of the wire potential for two different incident photon energies.

6 μm thick mylar (polyethylene-terephtalate).* The photon now loses its energy stepwise by ionizing argon atoms. The *average* energy required to ionize one argon atom is 25 eV, so that the

* Very thin windows are required for the detection of long-wavelength radiation. However, argon gas may leak through the window into the spectrometer. For this reason a continuous flow of gas is supplied to the tube, and, accordingly, such a detector is called a gas flow proportional detector.

average number of primary electrons produced by the photon is *proportional* to the photon energy. A photon with an energy of 4 keV (wavelength 0·31 nm) produces on the average 160 primary electrons.

These electrons are attracted towards a thin wire in the centre of the tube to which a positive voltage is applied. Fig. 8.6b shows the number of electrons arriving at the wire as a function of the positive potential of the wire. For potentials up to 100 V the primary electrons move very slowly and are immediately recaptured by the positive argon ions, so that no current pulse is measured. At slightly larger potentials only the primary electrons reach the wire, but their number is too small to be practically useful.

However, if the potential of the wire is raised to a value of about 1000 V, the primary electrons are strongly accelerated, and during their flight towards the wire they will liberate new electrons in collisions with argon atoms. The number of secondary electrons arriving at the wire is now several orders of magnitude greater, but this number is *still proportional* to the original energy of the incident X-ray photon (Fig. 8.6b). The potential of the wire now lies in the *proportional* region and, accordingly, the detector is called a proportional counter or, more correctly, a *proportional detector*.

If the potential is increased still further, the Geiger region is reached. The number of electrons arriving at the wire is now *independent* of the photon energy. The number of electrons is extremely large and the resulting current pulse is strong and can be measured without further amplification. Consequently, the Geiger-Müller detector is cheaper to use than the proportional detector. This type of detector is applied in the detection of β-radiation, but it is no longer used in X-ray spectrometry.

The proportional detector is perfectly suitable for X-rays with wavelengths exceeding 0·1 nm. Smaller wavelengths have correspondingly larger photon energies and are, therefore, less efficiently absorbed by the argon gas (the mass absorption coefficient decreases; see Fig. 8.3). For the detection of short-wavelength radiation (photon energy exceeding 15 keV) the *scintillation detector* is used, which is described in § 9.3. In this detector the X-ray photon is absorbed in a small crystal (2 mm thick NaI) which converts the photon energy into a light flash with a wavelength of about 400 nm. In turn, the intensity of the flash is converted into an electrical pulse by means of a photomultiplier tube (§ 4.3). Again, the magnitude of the current pulse is proportional to the energy of the incident X-ray photon.

While the electron avalanche due to one photon is being created and destroyed (with the light flash as an intermediary stage, in the scintillation detector), the detector is unable to detect other photons. During this *dead time* the slower positive ions are neutralized by a molecular gas (usually methane) added to the extent of about 10 per cent to the argon in the proportional detector. Typical values for the dead time are 300 μs for the Geiger-Müller detector, but only 0·5 μs for the proportional detector and 0·2 μs for the scintillation detector. The latter two detectors are thus capable of measuring an extremely rapid sequence of separate photons. The proportional detector and the scintillation detector are suitable for measuring high-intensity X-ray beams, and correctly indicate counting rates up to 100000 counts per second, whereas the Geiger-Müller tube already shows an error for counting rates as low as 1000 counts per second (although this may be corrected to some degree).

The number of electrical pulses produced by a beam of X-rays is actually a statistical quantity and subject to small random fluctuations. When T is the total number of counts measured during a certain counting time, the standard deviation that determines the random *counting error* is equal to $\sqrt{T}$. The relative random error is thus $1/\sqrt{T}$, and this can obviously be reduced indefinitely by increasing the counting time. An important advantage of the proportional detector and the scintillation detector is the large counting rate allowed. When the X-ray beam is sufficiently intense, the counting time can still be limited to a conveniently small value. For example, with a counting rate of 100000 counts per second the relative error is already reduced to 0·1 per cent for a counting time of only 10 seconds!

No doubt, the most valuable property of the proportional detector and the scintillation detector is the proportionality between the incident photon energy and the number of electrons produced. As a result, the magnitude of the electrical pulse or the *pulse height* is proportional to the photon energy. Usually, the detector receives a monochromatic beam of photons with a single energy (Fig. 8.7).

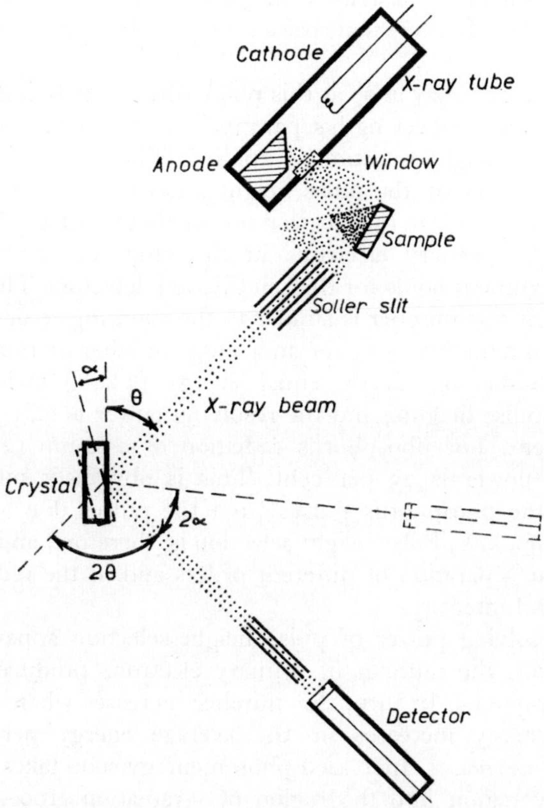

FIG. 8.7. Arrangement for X-ray fluorescence spectrometry.

The detector then yields a series of pulses of equal magnitude or of constant pulse height.

However, Equation 8.3 shows the possibility of overlapping orders. Phosphorus K_α-radiation of 0·616 nm will be diffracted at the same angle in the first order as calcium K_β-radiation of 0·309 nm in the second order. The X-ray beam then contains two wavelengths and photons of two different energies (namely 2 and 4 keV). The detector now yields current pulses of different magnitudes. These may be separated with a *pulse height selector*; this instrument is described in § 9.4, and may be considered as an electrical filter that passes only pulses of a preselected height.

The question may arise at this point why an analysing crystal is still retained for wavelength separation (that is, energy separation), when pulse height selection is applied. The answer is that the resolving power of the pulse height selector is rather small for X-rays. This is due to the fact that not all photons of 4 keV produce exactly 160 primary electrons in the proportional detector (a similar argument holds for the scintillation detector). The random variation of this number is similar to the counting error discussed above, and amounts to 13 for an average number of 160. Accordingly, photons of exactly equal energy (4 keV) yield slightly different pulse heights, and the resolving power is only 26/160 or 16 per cent. For phosphorus radiation of 0·62 nm (2 keV) the resolving power is 25 per cent. This is obviously sufficient to separate the phosphorus pulses from the pulses due to calcium photons of 4 keV. Pulse height selection is, therefore, an important tool in the separation of different orders and in the reduction of background intensity.

The resolving power of pulse height selection apparently increases with the number of primary electrons produced by the incident photons. In turn, this number increases when either the photon energy increases or the average energy per electron produced decreases. Increased photon energy soon takes us out of the X-ray region into the region of γ-radiation (100–2000 keV) where, indeed, pulse height selection is used exclusively to separate the photons of a γ-spectrum (§ 9.4). A few years ago, however, a new type of detector was developed for γ-spectrometry, in which the average energy needed to produce an electron is only 2·5 eV (§ 9.3). With this solid-state semiconductor detector the number of electrons produced by a 4 keV X-ray photon is raised to 1600, and the resolving power of pulse height selection is $2/\sqrt{1600}$ or about 5 per cent. This is still inferior to the resolving power of a good analysing crystal. At present, the possibilities of non-dispersive X-ray spectrometry (without analysing crystals) seem to be restricted to wavelengths below 0·05 nm (photon energies exceeding 25 keV).

8.5. X-ray fluorescence spectrometry

The optical arrangement of an X-ray fluorescence spectrometer is shown schematically in Fig. 8.7. Continuum radiation from the anode of the X-ray tube strikes the sample at an angle of about 60°. Excited fluorescence radiation is emitted in all directions, but only the beam leaving the sample at an angle of about 30° is selected. This is done by means of a *Soller slit* made of a number of parallel metal plates with a typical separation of 0·3 mm. Note that in contrast to the situation discussed in § 8.2 the primary X-ray beam is actually *polychromatic*. The various wavelengths of the secondary beam are separated by rotating the analysing crystal. As is clear from Fig. 8.7, the detector (a proportional or scintillation detector) must also be revolved around the centre of the analysing crystal and at twice the speed. The position of the detector is read from a precision goniometer that measures the diffraction angle θ or rather the angle 2θ between the incident beam and the diffracted beam. According to Equation 8.3 this angle is a measure of the wavelength of the detected X-rays. When the electrical signal from the detector is amplified and fed into a recorder the complete X-ray fluorescence spectrum can be scanned by slowly rotating the crystal and the detector.

The X-ray spectrum is essentially a line spectrum, the width of the lines being determined by the resolving power of the spectrometer. The line width is usually of the order of 0·3° on the 2θ scale, which means that the entire spectrum can contain only a few hundred lines (compare Fig. 8.8b). Fortunately, most elements show only two or three strong lines. However, overlapping lines are a fairly common phenomenon in X-ray spectrometry. If these lines have similar wavelength (not due to overlapping orders), pulse height selection offers no improvement. The resolving power can be increased by choosing an analysing crystal with smaller *d*-value or by polishing the crystal surface. However, a small *d*-value restricts the maximum wavelength that can be measured (§ 8.3), and a highly polished crystal shows poor reflectivity (actually, the crystal surface is purposely roughened to increase its reflectivity and hence the intensity of the peaks).

In principle, the X-ray spectrum is suitable for qualitative analysis, each line indicating the presence of a particular compound in the sample. However, detection limits in X-ray fluorescence spectrometry are not very low, and sample constituents below 0·01 per cent will not be noticed in a single scan of the spectrum.

The important application of X-ray fluorescence spectrometry is in quantitative analysis. The detector is adjusted to a single spectral line and kept in that position for a fixed counting time or for a fixed number of counts, T. As has been remarked in § 8.4, the relative random error is given by $1/\sqrt{T}$. For a counting rate of 5000 counts/s and a counting time of 3 minutes, the relative random error would be only 0·1 per cent. In practice, the instability of the X-ray tube and other parts of the instrument restrict the random error to a minimum value of about 0·2 per cent. However, this is still an extremely low value for an instrumental method of analysis, and fully comparable to wet chemical procedures (titrimetric and gravimetric analysis). As a result, X-ray fluorescence spectrometry is very suitable for the precise determination of major constituents in samples.

With respect to the systematic error the situation is less favourable. As was discussed in § 8.2, a difference in the absorption of X-rays between the sample and the standards used for preparing the analytical curve causes appreciable systematic errors. For the routine analysis of a great number of similar samples (as in production control) an obvious solution is to adapt the standards to the composition of the samples. In fact, a few typical samples may be analysed by other methods and used as standards in subsequent X-ray determinations. For samples of widely varying composition and for powdered samples in general, the borax fusion technique (compare § 7.5) is commonly applied to achieve constant absorption characteristics and to minimize particle size effects. A recent development is the application of digital computers to correct mathematically for the influence of interfering elements, and to take account of non-linear analytical curves etc.

According to the definition in § 1.6, the limit of detection is determined by the noise of the background. Since the random error (i.e. the noise) in measuring background intensities can be reduced at will by prolonging the counting time, it would seem that detection limits in X-ray spectrometry can be made arbitrarily low. Practical reasons and instrument drift restrict the counting times to several minutes. Consequently, limits of detection are of the order of 10^{-4} per cent (1 ppm) for elements with atomic number 20 to 40, and about 10^{-3} per cent (10 ppm) for elements with atomic number 40–90.

8.6. X-ray spectrometry with direct electron excitation

At several places in the previous sections the problems associated with the determination of light elements ($Z < 15$) have been indicated. These may be summarized as follows:

(*a*) The long-wavelength region most effective in exciting the *K*-lines of light elements is poorly represented in the continuum spectrum of the X-ray tube (Fig. 8.1), and is further reduced by absorption in the tube window.

(*b*) The *K*-lines of light elements are appreciably absorbed by the detector window and are less effectively detected.

(*c*) The fluorescence yield factor decreases rapidly with decreasing atomic number, the competitive process of Auger emission becoming more important.

(*d*) The depth of penetration of long-wavelength radiation into the sample is very small, so that the number of atoms contributing to the fluorescence spectrum is comparatively small.

In terms of Equation 8.3 this means that for light elements the intensity I_0 and the fluorescent yield Φ are small, whereas the mass absorption coefficients μ and μ' are very large. As a result I_F is small. Consequently, the detection limits cited at the end of the preceding section increase rapidly with decreasing atomic number. The determination of sodium and magnesium is already problematic and the determination of even lighter elements is not possible with conventional X-ray fluorescence spectrometers. Windowless X-ray tubes have been proposed for the determination of F, O, N and C down to 0·1 per cent, but these have not yet reached the stage of general applicability.

The intensity of excited X-rays can be markedly enhanced by

bombarding the sample with fast electrons. The sample then plays a role similar to the target in an X-ray tube. However, this technique introduces several problems. The entire spectrometer and the sample must be brought to the low-level vacuum required for the electron beam (10^{-7} atm.). The sample will be strongly heated by the incident electrons and the resulting volatilization will alter the X-ray intensities during the measurement. The characteristic lines are superimposed upon a strong continuum background spectrum, and good wavelength separation is imperative. Consequently, direct excitation with electrons is not an attractive alternative for the determination of light elements.

However, this form of X-ray spectrometry offers another opportunity. As a result of their charge, the electrons can be focused by electric and magnetic coils to an extremely fine spot (as is done in an electron microscope). In this way X-rays can be excited in sample areas as small as 1 μm^2. Typical applications of the *electron microprobe* are the identification of very small inclusions (1 μm^3) in steel samples and the analysis of metallic elements in single cells of biochemical samples.

8.7. X-ray diffractometry

Equation 8.4 states the unambiguous relation between the separation d of two adjacent atomic layers in a crystal, the wavelength λ of the X-rays and the reflection angle θ. In § 8.3 this relation was used to measure the wavelengths reflected consecutively by a crystal of known spacing d. In this section we consider the alternative possibility of deriving the crystal spacing d from the reflection angle θ if the crystal is irradiated with a monochromatic beam of X-rays.

In Fig. 8.5 it was assumed that the X-rays are reflected only by the atomic layers parallel to the surface of the crystal, but this choice is completely arbitrary. When the atoms of the crystal are connected (in thought) by other lines, different planes of atomic layers are obtained with correspondingly different separations d', d'', etc. With the crystal in the fixed position of Fig. 8.5, reflections against these other sets of planes cannot be observed, because Equation 8.4 is not obeyed. However, if the crystal, still being irradiated with monochromatic X-rays, is rotated around an axis perpendicular to Fig. 8.5, then each set of parallel planes will reflect the beam at a specific reflection angle. Indeed, Equation 8.4 shows that for a fixed wavelength λ, the spacings d, d', d'', etc. correspond to characteristic reflection angles θ, θ', θ'', etc.

Obviously, these separations can be expressed in the dimensions of the smallest unit of the crystal. In the two-dimensional Fig. 8.5 this *elementary cell* is the rectangle formed by four large neighbouring atoms. In an actual crystal the elementary cell is, of course, three-dimensional, and it need not be rectangular. Conversely, the dimensions of the elementary cell of a crystal and the location of the atoms in an elementary cell can be derived from the separations d, d', d'' etc., computed from the measured reflection angles θ, θ', θ'' etc. Admittedly, this is not an easy task, and high-speed digital computers are required to perform the complicated calculations. Nevertheless, *X-ray diffractometry* is a powerful tool in determining the spatial structure of crystals and molecules.

Within the context of this book we shall limit the discussion to the direct analytical application of X-ray diffractometry, which is the recognition and the quantitative determination of crystalline substances from their measured reflectance angles. If one single, fairly large crystal is used, it is necessary to rotate the crystal to observe the reflections from different planes. It is more convenient to grind the crystal to a fine powder, each grain forming a tiny crystal of its own. When a collection of grains is irradiated, there will always be some grains in a suitable position to reflect the beam at a particular crystal plane. As a result, the powder deflects the incident beam over various angles each corresponding to a particular d-value. This technique is called the Debije-Scherrer technique after its inventors.

Fig. 8.8a shows the arrangement of a diffractometer for powder analyses. A nickel filter absorbs the continuum radiation and the K_β-lines, and only the K_α-lines of the copper tube are transmitted (problem 7). The X-ray beam is made parallel by a Soller slit, and strikes the finely powdered sample contained in a thin tube accurately positioned in the centre of a circular camera. The major portion of the beam passes the sample unhindered and leaves the camera through a hole. The beams deflected symmetrically to both sides of the incident beam are detected on a photographic film

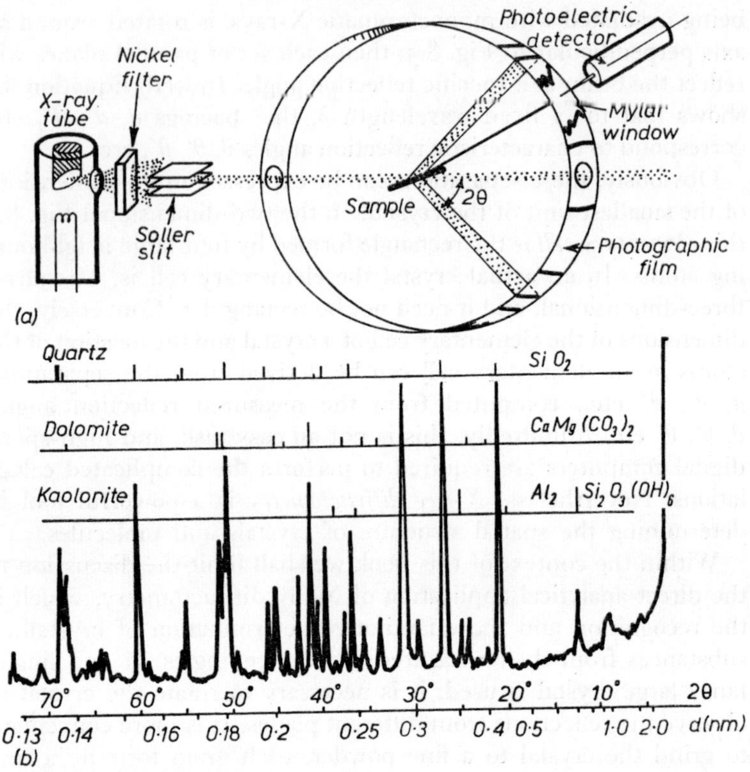

FIG. 8.8. (a) Arrangement for X-ray diffractometry according to the Debije-Scherrer technique. (b) X-ray diffraction spectrum of a mixture of equal proportions of three minerals recorded with an X-ray tube with copper anode, a nickel filter and a proportional detector.

attached to the inside of the camera circumference. Alternatively, the sample can be placed in the centre of a precision goniometer and the deflected beams can be recorded successively by a scanning proportional detector.

In both cases an X-ray diffraction pattern is obtained: a series of black lines on the photographic film, or with the proportional detector, a series of peaks on recorder chart paper. The distance of a line from the extremely strong central peak is a measure of the deflection angle θ (or more correctly 2θ, see Fig. 8.8a). Because θ

can be varied from 2 to 70° the *d*-values that can be measured with copper K_α-radiation of 0·154 nm range from 0·08 to 2·2 nm. As has been remarked in § 8.3, *d*-values smaller than 0·13 nm do not occur in practice. The example given in Fig. 8.8b illustrates how the observed *d*-values and their relative intensities can be used to recognize the crystalline substances in the sample. The intensity of a characteristic, strong reflection may be used to measure the concentration of a particular constituent.

Whereas other methods of inorganic analysis determine only individual elements or ions (with the possible exception of inorganic infra-red analysis, § 5.10), X-ray diffractometry enables us to recognize complete molecular structures. In addition, it is possible to distinguish between different crystalline modifications of a compound. Examples are the determination of TiO_2 as anatase or rutile, and iron as austenite or martensite. X-ray diffractometry can only be utilized with crystalline substances. Glasses yield no diffraction pattern, so that X-ray diffractometry provides a simple method to check whether a fusion procedure (§§ 7.5 and 8.5) has indeed eliminated all crystalline structures.

Bibliography

1. JENKINS, R., and DE VRIÈS, J. L., *Practical X-ray Spectrometry*, Philips Bibliotheek, Eindhoven, Netherlands, 1967.
2. BIRKS, L. S., *X-ray Spectrochemical Analysis*, Interscience, New York, 2nd ed., 1969.
3. LIEBHAFSKY, H. A., PFEIFFER, H. G., WINSLOW, E. H., and ZEMANY, P. D., *X-ray Absorption and Emission in Analytical Chemistry*, Wiley, New York, 1960.
4. BIRKS, L. S., *Electron Probe Microanalysis*, Wiley, New York, 1963.

Problems

1. X-rays with a wavelength of 0·10 nm strike a nickel plate at an angle of 60°. The density of nickel is 8·9 g/cm^3 and the mass absorption coefficient is 122 cm^2/g. Calculate the depth of penetration at which the intensity of the incident beam has decreased to 1 per cent. (Note: the depth of penetration is measured perpendicularly to the surface.)

2. The primary beam of problem 1 is used to excite nickel K_α-radiation of 0·165 nm. The secondary radiation leaves the sample at an angle of

30°, and the mass absorption coefficient is 61 cm^2/g. What proportion of the fluorescence radiation excited at the greatest depth of penetration will emerge from the sample?

3. Fig. 8.7 shows that for very small angles θ (that is for short-wavelength X-rays) the crystal will only partly intercept the X-ray beam. If the crystal is 6·5 cm long and the diameter of the X-ray beam is 1·25 cm, what will be the smallest angle θ for which the beam is still completely intercepted by the crystal surface?

4. Calculate the binding energies of the electrons in the K- and L-orbits of antimony from the position of the absorption edges. From this calculate the wavelength of the antimony K_α-lines.

5. The angle 2θ between the incident and the deflected beams in Fig. 8.8a can range from 0° to 180°. However, when copper K_α-radiation is used no first order lines are observed in the X-ray diffraction spectrum above $2\theta = 80°$. Explain this.

6. The density of argon gas at room temperature and normal pressure is 1·78 kg/m^3, and its mass absorption coefficient at 1 nm is 3·4 m^2/kg. Calculate the percentage of X-rays with a wavelength of 1 nm that will be absorbed in a proportional detector 20 cm long, and with an argon pressure of 0·1 atm.

7. In addition to copper K_α-radiation, the copper target of an X-ray tube also emits copper K_β-radiation. According to Fig. 8.3 a nickel filter can be used to absorb the K_β-radiation selectively. Use the data in Fig. 8.3 to calculate the thickness of the nickel filter required to absorb 99 per cent of the copper K_β-radiation. To what extent will the copper K_α-radiation be absorbed? (The density of nickel is 8·9 g/cm^3.)

8. For a series of aluminium standards the following net counting rates are measured:

Concentration Al (%)	*Countings rate* (*counts*/*s*)
1·0	155
1·3	190
2·0	298
2·5	365

The sample yields a net counting rate of 230 counts/s. Calculate the aluminium content of the sample.

9. How long must the sample in problem 8 be counted to reduce the relative standard deviation to 0·5 per cent?

10. The counting rate of the background in problem 8 is 100 counts per second. What is the noise for a counting time of one minute? If it is assumed that the analytical curve of problem 8 is valid down to very low concentrations, calculate the detection limit for the aluminium determination with a counting time of one minute.

CHAPTER NINE

ACTIVATION ANALYSIS

9.1. Radioactivity and the atomic nucleus

Natural radioactivity was discovered at the end of the last century, when it was observed that some elements at the end of the periodic table, notably uranium and radium, emit 'radiation' with the remarkable property that it darkens a photographic emulsion even through protective metallic sheets. On closer inspection the radiation appeared to consist of three distinct components:

(*a*) α-radiation, consisting of helium nuclei, which are rapidly stopped even in air;

(*b*) β-radiation, consisting of electrons of variable kinetic energy, which are stopped by glass or thin aluminium sheets;

(*c*) γ-radiation, being electromagnetic waves of extremely small wavelength (0·01 Å) and correspondingly high photon energy (1 MeV), which can only be stopped effectively by heavy blocks of lead.

These phenomena soon led to entirely new concepts of the structure of the atom. The subsequent explanation of atomic spectra in the ultra-violet and X-ray regions has been discussed in Chapters 2 and 8. In the present section and the next one, the atomic model will be extended to account for the phenomenon of radioactivity.

In the most simple atom, the hydrogen atom, nearly all of the mass is concentrated in a small, positively charged particle, called a *proton*. This proton, with a mass of $1{\cdot}7 . 10^{-27}$ kg and a diameter of about 10^{-15} m or 10^{-5} Å, forms the nucleus of the hydrogen atom. At the relatively large distance of about 1 Å, one electron revolves around the proton, and the hydrogen atom as a whole is electrically neutral. Heavier atoms contain more electrons, and their nuclei contain a correspondingly larger number of protons. This number (Z) determines the position of the element in the periodic table.

However, the atomic mass increases more rapidly than would be predicted from the number of protons, because the atomic nucleus also contains neutral particles, which are called *neutrons*. The mass of a neutron is nearly equal to the mass of a proton and both particles are about two thousand times heavier than an electron ($m = 9.10^{-31}$ kg). For most elements *isotopes* are known which contain different numbers of neutrons in addition to a fixed number of protons in the nucleus. For example, one quarter of the natural chlorine atoms contain 17 protons and 20 neutrons, whereas the other three quarters contain 17 protons and 18 neutrons. Such isotopes are chemically indistinguishable, but in radioactive processes their behaviour is usually quite different.

Different isotopes of an element are distinguished by writing the total number of nuclear particles before the chemical symbol. The two chlorine isotopes are thus written as ^{35}Cl and ^{37}Cl (this is read as 'chlorine thirty-seven'). As the chemical symbol tells us the place of the element in the periodic table and hence its number of protons, this notation describes the atomic nucleus unambiguously. For convenience, the number of protons may be indicated separately, as follows: $^{37}_{17}Cl$. In lighter nuclei the numbers of protons and neutrons are nearly equal ($^{16}_{8}O$, $^{40}_{20}Ca$, $^{32}_{16}S$), but heavier nuclei contain more neutrons than protons. The three isotopes of natural uranium are $^{234}_{92}U$, $^{235}_{92}U$, and $^{238}_{92}U$, which contain respectively 142, 143 and 146 neutrons in addition to 92 protons.

However, neither of these uranium isotopes is stable, and their nuclei disintegrate by expulsion of a helium nucleus (α-radiation) consisting of two protons and two neutrons. As a result, other nuclei are formed; for example:

$$^{238}_{92}U \rightarrow ^{4}_{2}He + ^{234}_{90}Th \qquad (9.1)$$

The thorium isotope produced in this *nuclear disintegration* is also unstable, and the thorium nucleus emits an electron. This may be understood as the transformation of a neutron into a proton plus an electron, and the expulsion of the latter. The nuclear mass remains (nearly) the same, but the nuclear charge increases by one unit:

$$^{234}_{90}\mathrm{Th} \rightarrow {}^{234}_{91}\mathrm{Pa} + \mathrm{e}^- \tag{9.2}$$

In this way a whole chain of disintegration reactions occurs, through which $^{238}_{92}U$ is transformed in 14 steps into the stable lead isotope $^{206}_{82}Pb$.

Some of the nuclei involved in this reaction chain are formed in a state of high internal energy. The protons and neutrons may be thought to occupy different 'positions' in the atomic nucleus that correspond to different potential energies. In exactly the same way as was discussed for atoms and molecules in Chapter 2, an energy level diagram can be constructed for the atomic nucleus, which may drop from a level of high internal energy to a lower level by emission of electromagnetic radiation. This radiation is called γ-radiation. Radioactive decay by expulsion of either α-particles or β-particles is therefore usually attended by the emission of γ-radiation. Emission of electromagnetic radiation is normally almost instantaneous, and takes place within 10^{-14} s.

Radioactive decay by α- or β-emission is a random process. We cannot predict which particular nucleus will disintegrate next, but for a large number of similar nuclei we can state the fraction that will disintegrate within a certain period of time. The *decay rate*, the number of disintegrations per second, is proportional to the total number of radioactive nuclei present:

$$\frac{\mathrm{d}N}{\mathrm{d}t} = -\lambda N \tag{9.3}$$

where λ is the *decay constant*. After t seconds the original number of nuclei N_0 will be reduced to

$$N = N_0\, e^{-\lambda t} \tag{9.4}$$

The decay rate is commonly indicated by the *half-life* $t_{\frac{1}{2}}$; this is the time required to reduce the number of nuclei to one half the original value:

$$N(t = t_{\frac{1}{2}}) \equiv \tfrac{1}{2} N_0 = N_0\, e^{-\lambda t_{\frac{1}{2}}}$$

whence

$$\lambda t_{\frac{1}{2}} = \ln 2 = 0{\cdot}6931 \tag{9.5}$$

The half-life of the decay reaction 9.1 is $4{\cdot}5 \, . \, 10^9$ year! This large value explains why the uranium isotope of mass 238 still exists. Another isotope of uranium with mass 236 is 'already' reduced by half in 10^7 years and, since the earth has existed for about 10^9 years, this isotope is no longer present in natural uranium. The thorium isotope ^{234}Th of reaction 9.2 has a half-life of only 24 days and this isotope can only be found because it is continuously being created by reaction 9.1.

Each disintegrating uranium isotope-238 yields one helium nucleus. If we succeed in counting the number of helium nuclei, the decay rate in Equation 9.3 may be measured. From this the decay constant λ and the half-life $t_{\frac{1}{2}}$ can be deduced. Alternatively, the number of helium nuclei liberated per unit time is a measure of the number of uranium nuclei N. This is the basis of quantitative analysis by means of radioactive processes.

9.2. Artificial radioactivity and nuclear reactions

The number of naturally radioactive isotopes is limited. This is not surprising, because an isotope must have a half-life of at least 10^9 years to have survived the time of the earth's existence; shorter-lived isotopes have long since died out (unless they are still being formed, as is ^{234}Th). A large half-life means a small decay constant (Equation 9.5) and a correspondingly small decay rate (Equation 9.3). Hence all naturally radioactive isotopes emit very little radiation, and their use in analytical determinations is very limited.

However, it is possible to make radioactive isotopes artificially. By bombarding stable atomic nuclei with small particles such as helium nuclei (α), protons (p) and neutrons (n) *nuclear reactions* can be induced and radioactive products may be formed. Historically, the earliest example is the bombardment of aluminium with helium nuclei (in 1934):

$$^{27}_{13}\mathrm{Al} + ^{4}_{2}\mathrm{He} \rightarrow ^{30}_{15}\mathrm{P} + ^{1}_{0}\mathrm{n}$$

$$\downarrow t_{\frac{1}{2}} = 2{\cdot}5 \text{ min}$$

$$^{30}_{14}\mathrm{Si} + e^{+}$$

In the first instance this reaction produces neutrons and phosphorus nuclei; the latter are unstable and are rapidly converted into silicon nuclei by the ejection of a particle with the mass of an electron but with a *positive* unit of charge (a *positron*). The bombardment of aluminium with helium nuclei leading to the expulsion of neutrons is called an (α, n)-reaction. The above reaction may be abbreviated as:

$$^{27}\text{Al}(\alpha,\text{n})\,^{30}\text{P}$$

Since a positively charged helium nucleus and a positively charged aluminium nucleus mutually repel each other, according to Coulomb's law, the α-particle must have a high kinetic energy to approach the aluminium nucleus close enough to let a nuclear reaction occur. This kinetic energy is carried away by the neutrons, which leave the reaction with an energy of several MeV. Such high-energy neutrons are called *fast neutrons.*

Because neutrons carry no charge, much less energy is required to let a neutron react with an atomic nucleus. Indeed, it appears that slow neutrons with a small kinetic energy can produce nuclear reactions. Of well-known general interest are the *nuclear fission* reactions induced in heavy nuclei. Upon irradiation with slow neutrons the nucleus of the uranium isotope 235 is split into two large fractions with nuclear masses ranging from 70 to 160. Since these lighter nuclei between them contain fewer neutrons than the original nucleus, nuclear fission also produces neutrons. Whereas each fission requires one neutron, two or three neutrons are liberated.

These secondary neutrons have a high kinetic energy (fast neutrons), and are first slowed down in a *moderator* (water or graphite). The fast neutrons lose energy through collisions until their kinetic energy is reduced to that of gaseous molecules at room temperature. The energy of such *thermal neutrons* is of the order of kT, where k is Boltzmann's constant and T is about 300°K (room temperature). Accordingly, thermal neutrons have a kinetic energy of $4 . 10^{-21}$ J or 0·025 eV.

The resulting slow neutrons can be used for further fission reactions. This leads to a self-sustaining reaction which, in addition to various nuclei and neutrons (fall-out) generates a large amount of energy. The energy yield can be calculated from the mass difference between the reacting particles and the reaction products, with Einstein's relation $E = mc^2$ (§ 9.3). Nuclear fission reactions are the source of nuclear energy in atomic bombs and *nuclear reactors*.*

However, thermal neutrons can also be used for analytical purposes. Many atomic nuclei are capable of accepting a neutron. Sometimes this leads to (n, α)- or (n, p)-reactions, but in the majority of cases an isotope of the same element is formed, with a mass increase of one unit. The excess energy of the reaction is lost very rapidly by emission of a γ-photon. Accordingly, this is known as an (n, γ)-reaction.

TABLE 9.1. Examples of neutron activation analysis

Isotope used	*Isotopic fraction*	*Cross-section (in 10^{-28} m²)*	*γ-photons (in MeV)*	*Half-life*	*Limit of detection* (in ng)*
^{107}Ag	0·51	45	0·6; 0·4	2·4 min	0·01
^{109}Ag	0·49	3·2	0·66; 0·88	250 days	1000
^{27}Al	1	0·24	1·78	2 min	0·2
^{140}Ce	0·88	0·31	0·145	33 days	1000
^{142}Ce	0·11	0·94	0·29	33 hours	100
^{59}Co	1	20	1·17; 1·33	5 years	300
^{58}Fe	0·003	1·01	1·10; 1·29	45 days	$5 \cdot 10^4$
^{127}I	1	5·6	0·45	25 min	0·05
^{115}In	0·96	150	0·40	54 min	0·003
^{23}Na	1	0·54	1·37; 2·75	15 hours	1
^{121}Sb	0·57	7	0·56	2·7 days	5
^{186}W	0·28	34	0·48; 0·69	1 day	1

* Limit of detection calculated as the amount of element that yields one count per second after an irradiation time of one hour, with a neutron flux of 10^{17} per m² per s and a counting efficiency of 10 per cent.

Quite frequently the isotope formed by an (n, γ)-reaction is unstable and decays into another atomic nucleus by ejection or by capture of an electron. As was remarked in the preceding section such decay reactions are usually accompanied by γ-emission. The energy of these γ-photons varies roughly from 0·1 to 3 MeV, as is shown by the examples in Table 9.1. As an example let us consider the irradiation of sodium with thermal neutrons (Fig. 9.1):

$$^{23}_{11}Na + ^{1}_{0}n \rightarrow ^{24}_{11}Na + \gamma$$

* Even more energy is produced in nuclear *fusion* reactions where the nuclei of very light elements are fused together to form heavier nuclei. One unfortunate application of this process is the hydrogen bomb. So far, it has not been possible to tame this reaction for use in nuclear fusion reactors.

The primary result of the *activation reaction* is the formation of ^{24}Na in a state of high energy, from which it recedes instantaneously to the ground state by emission of a γ-photon. However ^{24}Na is unstable and decays by electron emission into the most abundant magnesium isotope ^{24}Mg. The half-life of this decay reaction is 15 hours, and hence the decay reaction continues long after the irradiation has been stopped. The magnesium isotope is formed in a state of high energy from which it returns stepwise to the ground state by successive emission of two γ-photons. Since γ-emission is practically instantaneous this means that each electron ejected by a nucleus of sodium-24 is accompanied by two γ-photons of 2·75 and 1·37 MeV. The sodium isotope (and hence the original ^{23}Na present in the irradiated sample) may be recognized by its electron emission as well as by the γ-radiation from the excited ^{24}Mg. In fact, the latter is much more convenient.

Similar processes occur after many (n, γ)-reactions. At present, 70 elements can be detected through radioactive isotopes formed in (n, γ)-reactions. This forms the basis of *neutron activation analysis*.

9.3. Detection of γ-rays

Upon irradiation of a sample with thermal neutrons many elements form radioactive isotopes through (n, γ)-reactions. In the radioactive decay of each isotope one or more characteristic γ-rays are emitted. The wavelength or the *photon energy* of a particular γ-ray is a qualitative indication of the presence of a particular element in the sample. The *rate* of photon emission is a quantitative measure of the amount of that element (§ 9.5). Both qualitative and quantitative applications of γ-spectrometry are possible once we succeed in obtaining a γ-spectrum with fixed positions allotted to the γ-photons of each particular energy.

Unfortunately, for γ-rays there is no known dispersive device equivalent to the prism or the grating for optical radiation, or to the analysing crystal for X-rays. Consequently, γ-rays cannot be deviated into different directions according to their wavelength. However, a γ-spectrum may be obtained in another way, by the application of *pulse height selection*, the principle of which has been discussed in § 8.4. Just as in X-ray spectrometry, the detector

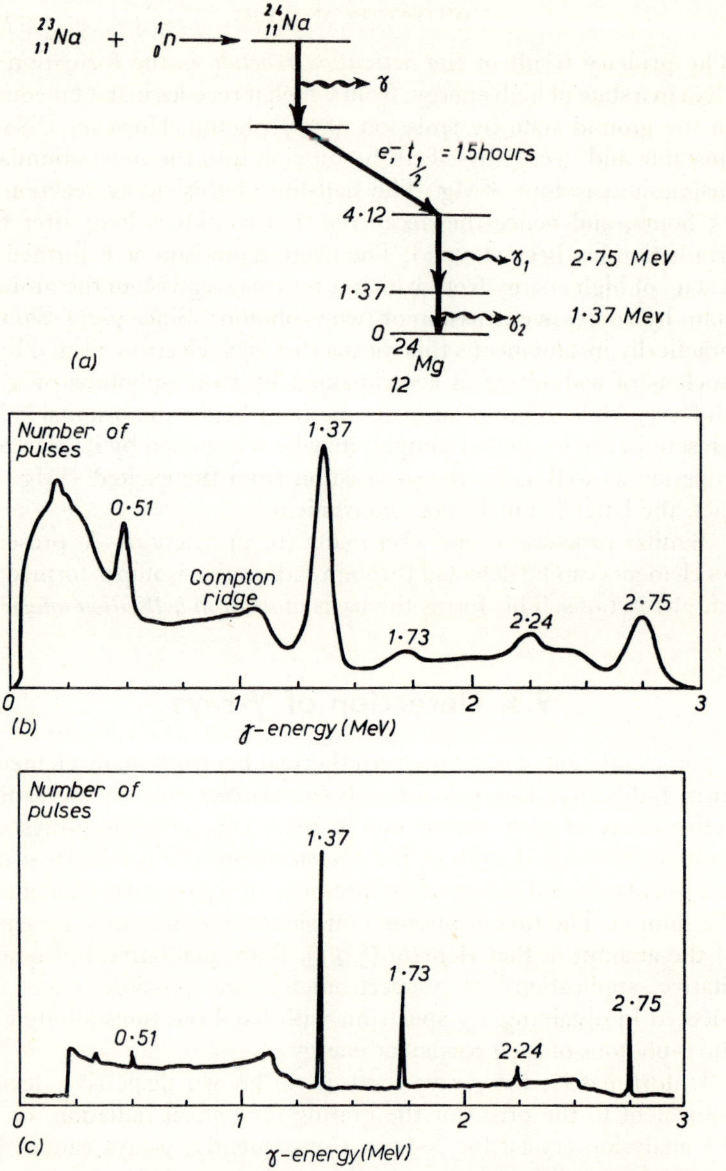

FIG. 9.1. Gamma-spectrum of ^{24}Na. (a) Activation of ^{23}Na by thermal neutrons, and decay scheme of ^{24}Na. (b) Gamma-spectrum of ^{24}Na recorded with a scintillation detector. (c) Gamma-spectrum of ^{24}Na recorded with a semiconductor detector. Note that we speak of the γ-spectrum of sodium, although, strictly speaking, the spectrum results from excited ^{24}Mg nuclei. (The kindness of Drs. W. M. C. de Jong, Reactor Institute, Delft, in providing the spectrum is gratefully acknowledged.)

employed in γ-spectrometry converts the energy of a single photon to an electric pulse with an amplitude (the pulse height) proportional to the photon energy. The electric pulses are selected and collected according to their pulse height, and a spectrum of electric pulses is obtained that is equivalent to the spectrum of γ-rays emitted by the sample.

However, this procedure is complicated by the fact that γ-radiation of a single photon energy gives rise to electric pulses of various heights. Mono-energetic γ-radiation does not yield one single peak, but in itself produces an entire γ-spectrum (see Fig. 9.1). The reason for this may be understood by considering in some detail the processes that occur when γ-rays interact with matter.

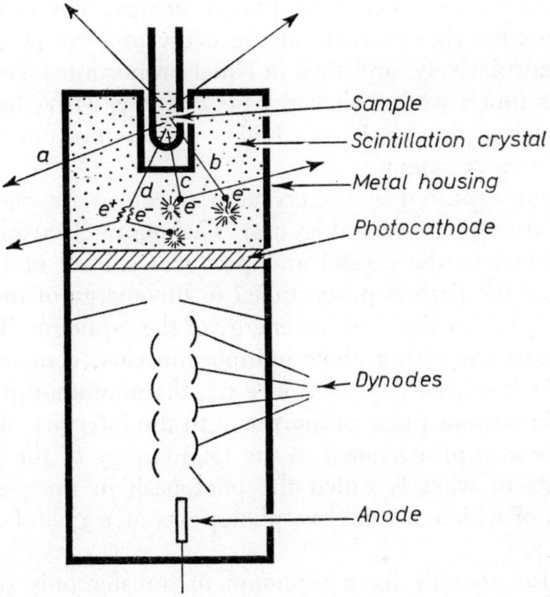

FIG. 9.2. The scintillation detector. The arrows *a*, *b*, *c* and *d* represent γ-photons interacting in four different ways with the scintillation crystal. *a*. The photon penetrates the crystal entirely and is not detected. *b*. The photon transfers its entire energy to an electron and gives a contribution to the photopeak. *c*. Compton-scattering; a photon of reduced energy leaves the crystal undetected. *d*. Pair-production followed by annihilation radiation.

Fig. 9.2 shows schematically the arrangement of a *scintillation detector*, which is still the most widely used detector in γ-spectrometry. The solid or dissolved sample is placed into a cylindrical hole drilled into a large sodium iodide crystal doped with 1 per cent thallium. The crystal is enclosed by a light-tight metal case. The arrangement ensures that a large fraction of the emitted γ-rays reaches the crystal. This is important, because high energy γ-photons have a large penetrating power and quite a few of the photons simply traverse the crystal without being detected at all (like photon *a* in Fig. 9.2). The efficiency of γ-capture increases with increasing crystal size and with increasing nuclear charge of the crystal elements. Therefore, large (100–300 cm^3) iodide ($Z = 53$) crystals are used. However, the efficiency of γ-capture also decreases with increasing photon energy. For example, the two γ-photons characteristic of the decay of ^{24}Na (Fig. 9.1) are emitted successively, and thus in equal proportions. Yet the peak at 2·75 is much weaker than the peak at 1·37 MeV, because the high energy γ-photons have a higher penetrating power and are lost to a greater extent.

A photon captured by the crystal may convey its whole energy to an electron (photon *b* in Fig. 9.2). This photo-electron is rapidly slowed down in the crystal and produces a flash of light. The *intensity* of the flash is proportional to the energy of the electron and thus is proportional to the energy of the γ-photon. The intensity is measured with a photomultiplier in close contact with the crystal. As has been described in § 4.3, the photomultiplier yields an electric current pulse proportional to the intensity of the flash and hence also proportional to the total energy of the γ-photon. This leads to what is called the photopeak in the γ-spectrum, examples of which are the two photopeaks at 1·37 and 2·75 MeV in Fig. 9.1.

It is also possible for a γ-photon to transfer only part of its energy to an electron in the crystal, by a process known as *Compton scattering* (photon *c* in Fig. 9.2). In this case if the energetically weakened photon then leaves the crystal unhindered, the smaller energy of the electron gives rise to a light flash of reduced intensity, and this produces in the photomultiplier a smaller current pulse which contributes to the lower energy side of the spectrum. Consequently, each photopeak in the γ-spectrum is attended by a

Compton ridge at a lower energy that ranges over a fairly large region of the spectrum and forms an inconvenient background to lower energy peaks (Fig. 9.3b).

So far, the explanation of the γ-spectrum is fairly straightforward. However, some additional sharp peaks may be noted in the spectra of Fig. 9.1 which cannot be traced directly to the decay scheme of ^{24}Mg. These peaks arise basically from the equivalence of energy and matter expressed by Einstein's equation, $E = mc^2$. According to this expression the mass of an electron corresponds to an energy of $8{\cdot}2 . 10^{-14}$ J or 0·51 MeV.

Now, a γ-photon with an energy exceeding 1·02 MeV may give rise to the spontaneous creation of an electron (e^-) and a positron (e^+). Since electrical neutrality must be maintained, the two particles can only be created simultaneously, and this process is known as *pair-production* (photon *d*).

The energy of the photon is thus decreased by 1·02 MeV. Within a very short period of time the positron will combine with an electron to form *two* γ-photons each with an energy of 0·51 MeV. So, instead of the original single photon, there are now three photons present in the crystal. If all three of these are captured by the crystal the resulting *total* light flash again contributes to the photopeak, because it is clear that none of the original photon energy is lost. However, it is also possible that one or two of the three photons traverse the whole crystal, like photon *a*. As a result, we observe in the spectrum of sodium additional peaks at an energy of $2{\cdot}75 - 1{\cdot}02 = 1{\cdot}73$ MeV (*escape peak*), and at 0·51 MeV and 1·02 MeV (*annihilation peaks*).

In this way a γ-spectrum consists of photopeaks, Compton ridges, escape peaks and annihilation peaks. The relative intensity of the peaks depends mainly upon the energy of the γ-photons and upon the size of the crystal. A low-energy γ-photon is almost completely converted into a photopeak in the scintillation detector; high-energy photons contribute only about 20 per cent of their energy to the photopeak.

The advantages of the scintillation detector are its fairly large efficiency and its relatively low price. However, a major disadvantage of the scintillation detector is its low resolving power, which is due to the appreciable width of individual peaks. The half-width (the width of a peak at half its maximum intensity) amounts to about 0·05 MeV. Since even a single isotope gives rise to a fairly complicated spectrum (Fig. 9.1b), the γ-spectra of a mixture of different radioactive isotopes will strongly overlap. Therefore, the use of a scintillation detector will frequently require preliminary separation of radioactive species by chemical procedures (§ 9.6).

A much better resolving power is obtained with a *semiconductor* detector made of germanium doped with a little lithium. Again, the γ-photon transfers its energy partly or completely to an electron, which produces further ionization within the detector. The germanium becomes slightly conductive, and the electrons can be collected into a current pulse under the influence of an applied electric potential difference. Unfortunately, the current pulses are very much weaker than the pulses produced by the photomultiplier of a scintillation detector. Accordingly, highly sensitive, low-noise electronic instruments must be used to amplify the signals, and the semiconductor detector must be cooled under vacuum with liquid nitrogen further to reduce the noise. An important disadvantage of this detector is its low efficiency (0·1 per cent), because it has not yet been possible to make homogeneous semiconductors of large size. However, there is rapid progress in this field, and already semiconductors with an effective volume of up to 30 cm^3 are available.

As compared to the scintillation detector, the resolving power of the semiconductor detector is at least ten times better, and accordingly the γ-spectrum now consists of a series of sharp peaks superimposed upon a smooth background (Compton ridge), as is shown by the spectrum in Fig. 9.1c. The detailed spectra obtained with this detector allow the identification and determination of several elements simultaneously. Preliminary chemical separations are then greatly simplified or even rendered unnecessary. In the latter case, we speak of *non-destructive* analysis.

9.4. The γ-spectrometer

The two main instruments required for neutron activation analysis are a source of thermal neutrons to activate the sample and a spectrometer to measure the γ-spectrum of the irradiated sample.

The most widely used source of thermal neutrons is the nuclear reactor, which offers the important advantage of high sensitivity and low limits of detection (§ 9.5) as a result of its very high neutron flux density (the number of neutrons passing one square metre per second). A considerable drawback of the nuclear reactor is its very high price. A small-sized nuclear reactor used for scientific purposes only still costs around one million dollars, so that obviously these cannot be built in large numbers.*

The analyst wishing to utilize neutron activation analysis then has two options. On the one hand he may send his sample to the analytical laboratory to be found near almost every nuclear reactor, and wait for the results. On the other hand he may have his sample irradiated for a specified period of time in a far-away reactor, and measure the γ-spectrum of the returned sample himself. For this he needs a laboratory suitable for handling radioactive materials (with all its safety requirements !), as well as a γ-spectrometer.

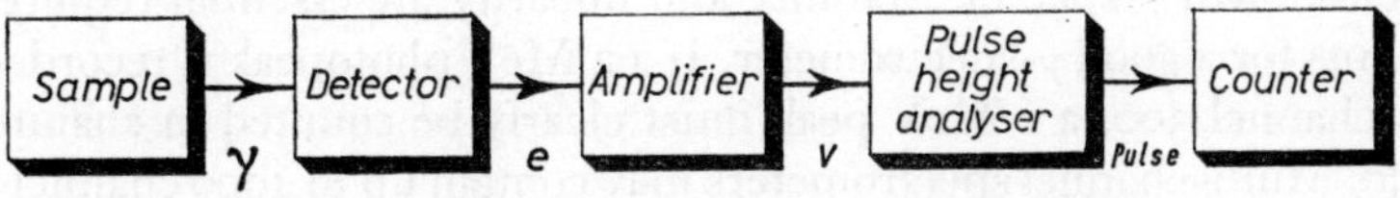

FIG. 9.3. Block diagram of a γ-spectrometer.

Fig. 9.3 is a block diagram of a γ-spectrometer. A γ-photon emitted by the sample generates an avalanche of electrons in the detector, that constitute a small current pulse. The current pulse is converted to a potential pulse and amplified to a potential difference of the order of a few volts. The pulses produced by the detector are then sorted according to their size by a *pulse height analyser* consisting basically of two electrical filters. Suppose, for example, that the first filter is adjusted in such a way that it passes only those (amplified) pulses exceeding a potential of 4·00 V, and suppose the second filter passes only those pulses exceeding 4·02 V. Pulses that are stopped or transmitted by *both* filters are

* Mini-sized nuclear reactors for neutron activation analysis only may be purchased for one tenth of this price, but their neutron flux density is about two orders of magnitude smaller.

not recorded, but any pulse transmitted by the lower filter but stopped by the higher filter is noted by a counter, so that the count increases by one unit. The two filters then determine a *channel* set for 4 V with a width of 0·02 V.

With a single-channel spectrometer the complete γ-spectrum may be scanned sequentially by raising the lower level of the filters, or alternatively by increasing the pulse amplification stepwise. In both cases pulses of increasing magnitude are recorded consecutively. The more expensive multi-channel spectrometer automatically converts each pulse to a count at the correct location in the spectrum. In this case, the complete γ-spectrum is recorded simultaneously, which is important for the study of short-lived isotopes. The spectrum may be displayed on an oscilloscope or, after some electronic operations, as a series of numbers denoting the number of counts in the various channels.

Obviously, the spectrometer must first be calibrated. The channel-number must be correlated with the photon energy of the γ-radiation. This is done with one or two radioactive isotopes with well-known γ-spectra. Stability and linearity are essential requirements for a good γ-spectrometer. If a 2 MeV photopeak is recorded in channel 300, a 1 MeV peak must clearly be counted in channel 150. Multi-channel spectrometers may contain up to 4000 channels.

9.5. The activity of an irradiated sample

If the γ-spectrometer is set to record the γ-photons of a certain photon energy (preferably a photopeak) and is operated for a fixed time, the total number of counts, divided by the counting time, is called the *counting rate* A. The counting rate is usually expressed in counts per second, and, as was remarked in § 9.3, the counting rate is a quantitative measure of the amount of the γ-emitting species in the sample.

A quantitative expression for the count rate is derived as follows. Let the sample contain N_Q atoms of an element Q, a fraction p of which are isotopes that can be converted into radioactive isotopes R through an (n, γ)-reaction. The amount of R formed per second is proportional to pN_Q and to the intensity Φ of the neutron flux. The proportionality constant σ_Q is called the *cross-section* of the activation reaction, and it denotes the readiness of the nuclei Q to react with neutrons to form the nuclei R. However, during irradiation the radioactive isotopes R will partly decay to the stable product S, according to Equation 9.3. Consequently, the net production rate of radioactive isotopes R is given by

$$\frac{dN_R}{dt} = pN_Q \Phi \sigma_Q - \lambda_R N_R \qquad (9.6)$$

The number of radioactive nuclei R produced during an irradiation time t_i is then given by

$$N_R = \frac{pN_Q\Phi\sigma_Q}{\lambda_R}[1-\exp(-\lambda_R t_i)]$$

$$= \frac{pN_Q\Phi\sigma_Q}{\lambda_R}[1-\exp(-0{\cdot}693\, t_i/t_{\frac{1}{2},R})] \qquad (9.7)$$

where Equation 9.5 has been applied.

After irradiation, the radioactive isotope R continues to decay according to Equation 9.3. Each disintegration produces one or more γ-photons. However, not all photons are detected, and not all detected photons yield pulses of the same size. Consequently, the count rate measured at a fixed position in the spectrum (e.g. one of the photopeaks) is only a fraction ϵ of the total decay rate. The magnitude of ϵ depends upon the decay scheme of the radioactive isotope R (Fig. 9.1a) and upon the efficiency of the detector (§ 9.3). Combination of Equations 9.3 and 9.7 yields for the count rate:

$$A_R = -\epsilon\frac{dN_R}{dt} = \epsilon\lambda_R N_R$$

$$= \epsilon pN_Q\Phi\sigma_Q[1-\exp(-0{\cdot}693\, t_i/t_{\frac{1}{2},R})] \qquad (9.8)$$

If the irradiation time t_i is *much smaller* than the half-life $t_{\frac{1}{2},R}$, this can be written as

$$A_R(t_i \ll t_{\frac{1}{2},R}) = \frac{0{\cdot}693\, \epsilon p N_Q \Phi \sigma_Q t_i}{t_{\frac{1}{2},R}} \qquad (9.9)$$

For long-lived isotopes R the count rate is inversely proportional to the (long) half-life $t_{\frac{1}{2},R}$, and this reduces the sensitivity of the analysis (compare Table 9.1, p. 194). However, in this case the count rate is also proportional to the irradiation time t_i, and elements yielding long-lived isotopes R may be irradiated for several days to increase the sensitivity of the analysis.

If the irradiation time t_i is *much larger* than the half-life $t_{\frac{1}{2},R}$, the exponential term in Equation 9.8 becomes very small, and

$$A_R\,(t_i \gg t_{\frac{1}{2},R}) = \epsilon p N_Q \Phi \sigma_Q \tag{9.10}$$

where now the count rate has become independent of the irradiation time! In practice, this means that it is pointless to irradiate an element for more than four times the half-life of the radioactive product. For example, if the radioactive product has a half-life of one hour (^{115}In in Table 9.1) the count rate levels off to a constant value after four hours irradiation, and does not increase significantly after prolonged irradiation.

Let us return now to Equation 9.8, to estimate the lower limit of detection that can be reached in neutron activation analysis. As an example, we consider the element sodium, for which the γ-spectrum is shown in Fig. 9.1. Natural sodium consists of a single isotope (^{23}Na), so that the fraction p is equal to one. The neutron flux available in a nuclear reactor is of the order of 10^{17} neutrons per m^2 per second, and the cross-section for thermal neutrons is given in the literature as $0{\cdot}54 \,.\, 10^{-28}$ m^2. If the sample is irradiated for one half-life (15 hours), the factor between brackets in Equation 9.8 is equal to 0·5. Finally, the efficiency factor ϵ may be estimated as 10 per cent, which means that one out of ten γ-photons emitted by the radioactive product (actually ^{24}Mg; see Fig. 9.1) contributes to the photopeak at 1·37 MeV. The count rate is then calculated as $A_R = 2{\cdot}7 \,.\, 10^{-13}\, N_Q$ counts per second.

In practice, it appears to be possible to distinguish a count rate of one count per second from the natural background due to cosmic rays. Hence, the limit of detection is about $3{\cdot}7 \,.\, 10^{12}$ atoms Q. For sodium, with atomic weight 23, this corresponds to $1{\cdot}4 \,.\, 10^{-10}$ g or 0·14 ng! In a sample of 1 g, sodium may be detected down to a concentration of 10^{-4} ppm (under favourable conditions). As is clear from the data in Table 9.1, however, the limit of detection becomes poorer, if the irradiation time is reduced to one hour. This agrees with Equation 9.9, and for such long-lived isotopes as ^{59}Fe and ^{60}Co an irradiation time of a few hours is too short for optimum sensitivity.

Whereas sodium consists of only one isotope, ^{23}Na, most other elements are composed of several isotopes. Sometimes the most abundant isotope cannot be activated by thermal neutrons. For example, natural iron consists of 92 per cent ^{56}Fe, which cannot be activated by thermal neutrons. Therefore, neutron activation analysis must be based upon ^{58}Fe, which is present to an extent of only 0·3 per cent in natural iron. Obviously, this small value for the isotopic fraction p raises the limit of detection.

Table 9.1 shows two examples where different isotopes of an element may be used for activation analysis. The case of cerium demonstrates that, again, the most abundant isotope need not be the best choice. ^{140}Ce is the most abundant isotope, but both the cross-section and the half-life of the radioactive product are markedly better for ^{142}Ce. As a result, the sensitivity of the determination is improved if the less abundant isotope is used for activation analysis. An even larger difference is observed for silver, where the cross-section and the half-life are much more favourable for ^{107}Ag than for ^{109}Ag. By using the shorter-lived silver isotope the limit of detection is decreased by five orders of magnitude. However, it is clear that the ^{108}Ag formed decays very rapidly, and within an hour most of its activity has vanished. The sample must therefore be measured directly after irradiation. Alternatively, if the sample is allowed to stand for one hour after irradiation, the short-lived isotope dies out and only the activity from ^{109}Ag (actually ^{110}Ag) persists. In practice, such *cooling periods* are utilized to reduce the background of the γ-spectrum due to the decay of short-lived species.

9.6. Quantitative analysis

Quantitative applications of neutron activation analysis are based upon the linear relation between the count rate, A, and the amount of the responsible element in the sample, N_Q, expressed by Equation 9.8. This equation cannot be used in an absolute way, because the values of ϵ, Φ and σ_Q are only approximately known and, moreover, the neutron flux varies with time and with the position in the nuclear reactor. Therefore, quantitative determinations are performed in a relative way by irradiating simultaneously a sample and a standard of known composition. The counting rates of the activated substances are measured under the same conditions, and the absolute amount of an element in the sample is calculated as

$$G_n = \frac{A_n}{A_{st}} G_{st} \tag{9.11}$$

where A_n and A_{st} are the counting rates for sample and standard, respectively, and G_{st} is the amount of the element present in the standard.

Several complications may give rise to systematic errors. Usually, the neutron flux is constant over the entire depth of the sample, but sometimes a major component with a large cross-section for thermal neutrons (such as cadmium) may weaken the neutron flux noticeably. It is then necessary to adjust the composition of the standard to that of the sample. It has also been remarked that the γ-spectrum of an activated sample may show many peaks. In the majority of cases the resolving power of the scintillation detector is insufficient to isolate the peak of interest, and preliminary separation of the desired component will be required. Precipitation, solvent extraction and ion exchange are the techniques most frequently applied. Further improvements in and the more widespread application of high-resolution semiconductor detectors will reduce the need for chemical pretreatment of irradiated samples. Even then it is not a simple matter to resolve the complex γ-spectrum into its various components, but fortunately computers can be used to perform this task.

Another potential source of systematic errors is unwanted reactions due to the occasional fast neutrons that are always present in the flux of thermal neutrons produced by the nuclear reactor. For example, if the sample contains copper and zinc, the determination of copper through the activation reaction with thermal neutrons $^{63}Cu\ (n, \gamma)\ ^{64}Cu$ will be perturbed by the presence of fast neutrons which react with zinc atoms according to $^{64}Zn\ (n, p)\ ^{64}Cu$. This interference cannot be removed by separation after the irradiation, but only by separation of the zinc from the copper prior to irradiation.

The precision and the limit of detection in neutron activation analysis are normally limited by the level of background activity which is notably due to the Compton ridge of the major component in the sample. The standard deviation in a measured number of T counts is equal to $\sqrt{T}$ (compare § 8.4). A strong background thus raises the limit of detection, which is given as $3\sqrt{T_b}$ (§ 1.6). Also, the precision in measuring a weak photopeak superimposed upon a strong background is obviously small. Again, the background intensity may be reduced appreciably by separation procedures after irradiation. Also, the background activity from short-lived isotopes can be reduced quite simply by letting the activated sample stand (cool) for some time before measuring the γ-spectrum. The non-uniform neutron flux in the nuclear reactor and the background intensity of the γ-spectrum limit the precision of quantitative neutron activation analysis to 2 per cent in favourable cases, and more typically to 5 per cent.

However, the disadvantages of complicated equipment (such as the nuclear reactor !), time-consuming irradiation and chemical treatment, and the only moderate precision, must be weighed against the distinct advantage of the very low limits of detection that can often be reached in neutron activation analysis. Activation analysis is extremely suitable for trace analysis or in the determination of microsamples. One advantage over other methods of trace analysis is that contamination of the sample *after the irradiation* introduces no errors, since such contaminations are not radioactive. Typical fields of application are biochemistry, clinical chemistry, analysis of semiconductors, forensic science, water and air pollution etc. Some elements, notably the rare earths, can only be determined with high sensitivity by neutron activation analysis.

9.7. Application of fast neutrons

Although most elements form radioactive products upon irradiation with thermal neutrons, there are a few elements for which this is not a competitive method of analysis (Be, Mg), or downright impossible. The very interesting elements oxygen, carbon and nitrogen cannot be activated by thermal neutrons, either because the cross-section of the reaction is very small or because the resulting isotope is not radioactive. Currently, interest is growing in determining these elements through activation by fast neutrons. Instead of a nuclear reactor, a *neutron generator* tube is used, in which a beam of high-energy deuterons (^{2}H) is directed towards a tritium target to yield fast neutrons with an energy of 14 MeV through the reaction $^{3}H(d, n)^{4}He$.

Neutron generators are small, and cheap enough to be installed in any laboratory, but the flux density of fast neutrons is only 10^{13} per m^2 per s, which lowers the sensitivity of the determination. Also, fast neutrons induce nuclear reactions of various kinds, which may sometimes interfere. For example, the fast neutron activation process $^{24}Mg\,(n, p)\,^{24}Na$ cannot be used to determine magnesium in rocks, because of the perturbing reaction $^{27}Al\,(n, \alpha)\,^{24}Na$. Therefore, the application of fast neutron activation will be reserved for a few cases, where other methods of analysis are less attractive. One important example is the determination of oxygen through the reaction $^{16}O\,(n, p)\,^{16}N$, down to one ppm of oxygen.

References

1. Friedlander, G., Kennedy, J. W., and Miller, J. M., *Nuclear and Radiochemistry*, Wiley, New York, 2nd edn., 1964.
2. Choppin, G. R., *Experimental Nuclear Chemistry*, Prentice-Hall, Englewood Cliffs, N.Y., 1961.
3. Lenihan, J. M. A., and Thomson, S. J. (eds.), *Activation Analysis*, Academic Press, London, 1965.
4. Duncan, J. E., and Cook, G. B., *Isotopes in Chemistry*, Clarendon Press, Oxford, 1968.
5. Perdijon, J., *L'analyse par Activation*, Masson, Paris, 1967.
6. Raković, M., *Activation Analysis*, Iliffe Books, London, 1970.
7. Crouthamel, C. E., Adams, F., and Adams, R., *Applied Gamma Ray Spectrometry*, Pergamon Press, 2nd edn., 1970.

Problems

1. What is the number of α-particles and β-particles emitted in the decay of one nucleus $^{238}_{92}U$ to the stable product $^{206}_{82}Pb$?

2. The naturally radioactive isotope with the shortest half-life is ^{40}K, with a half-life of $1{\cdot}27 . 10^9$ year; the proportion of this isotope in natural potassium is 0·012 per cent. Calculate the number of disintegrations per second in one gram KCl (Equation 9.3). Also, calculate the percentage of ^{40}K in natural potassium at the time the Earth was formed, $5 . 10^9$ years ago.

3. The exact masses of the neutron, the proton and the electron are

$$n = 1{\cdot}008665 \quad \text{AMU}$$
$$p = 1{\cdot}007277 \quad \text{AMU}$$
$$e = 0{\cdot}000549 \quad \text{AMU}$$

Use Einstein's equation to calculate the amount of energy liberated in the reaction

$$^1_0\text{n} \rightarrow {}^1_1\text{p} + \text{e}^-$$

4. Which isotopes will be formed in the first instance if $^{27}_{13}$Al reacts according to the following process?

$$(\alpha, n), (\alpha, p), (n, p), (n, \gamma), (p, n)$$

5. Check that differentiation of Equation 9.7 yields Equation 9.6.

6. The factor $1 - \exp(-0{\cdot}693\, t_i/t_{\frac{1}{2},R})$ in Equation 9.8 is called the saturation factor. Prepare a graph with the value of this factor plotted as a function of $t_i/t_{\frac{1}{2},R}$. Check the statement made in § 9.5 that there is no purpose in prolonging the irradiation for more than four times the half-life.

7. A 1 mg sample is irradiated for 25 min. with thermal neutrons. Immediately after the irradiation, the count rate for the photopeak of iodine is measured as 500 counts per second. Calculate the concentration of iodine in the sample, for an efficiency factor of 0·5 per cent and a neutron flux of 10^{16} per m^2 and per second (see also Table 9.1).

8. To what value will the count rate of the iodine photopeak of problem 7 be reduced if 90 minutes are lost by chemical separation of the iodine prior to the counting?

9. Explain the origin of the peaks at 0·35 and 2·24 MeV in the spectrum of Fig. 9.1c. Show that an infinitely large γ-detector will produce only photopeaks.

CHAPTER TEN

NUCLEAR MAGNETIC RESONANCE SPECTROMETRY

10.1. Nuclear spin

Around 1925 physicists studying the spectra of free atoms became aware that in addition to the strong electrical forces between the positively charged nucleus and the negatively charged electrons, much weaker forces must be present to account for the finer details in these spectra. These forces are of magnetic origin. The orbital motion of an electron around the nucleus constitutes a circular current, and this gives rise to a magnetic field in the same way as the circular windings of a solenoid or a transformer. It appears further that the electron itself and also some atomic nuclei behave like tiny magnets. This was first visualized by assuming that the electron revolves around its own axis (which would again constitute a small circular current) and the term 'electron spin' was introduced. However, this classical picture soon ran into difficulties when it was discovered that neutral particles can also have magnetic properties. A satisfactory and quantitative description of the magnetic forces inside an atom was developed in 1930 by Dirac on the basis of quantum mechanics. Nevertheless, the magnetic properties of electrons, protons, neutrons and atomic nuclei are still referred to as 'the spin' and characterized by a *spin quantum number* (§ 2.2).

Under normal conditions the magnetic properties of atomic nuclei are hardly noticeable, but this changes when the atoms are brought between the poles of a powerful magnet. The nuclear magnets will then respond to the influence of the magnetic field and strive for a favourable orientation, just as the needle of a compass will point in the direction of the north pole of the magnetic field of the Earth.

However, according to quantum mechanics there is an important difference between a compass needle and a magnetic atomic nucleus. If we apply some force with our finger, we can make the compass needle point in any direction, but the nuclear magnets can only have a limited number of orientations with respect to the magnetic field. This number of orientations can be derived from a *nuclear spin quantum number I*, as is illustrated in Fig. 10.1.

It has been found that atomic nuclei with an *even* number of protons *and* an *even* number of neutrons have no magnetic properties. Nuclei such as $^{12}_{6}C$, $^{16}_{8}O$, $^{32}_{16}S$, $^{86}_{36}Sr$ etc. have a nuclear spin $I = 0$ and do not react to the presence of a magnetic field: their energy remains unaltered.

The nucleus of the most abundant hydrogen isotope ($^{1}_{1}H$) consists of a single proton with a nuclear spin $I = \frac{1}{2}$. The proton now has *two* possible orientations in a magnetic field. One orientation is denoted by $m_I = +I = +\frac{1}{2}$. The proton-magnet is now inclined towards, but not exactly parallel to the magnetic field strength H, and this *lowers* the energy of the hydrogen nucleus. The other orientation is denoted by $m_I = -I = -\frac{1}{2}$. The proton-magnet now points in a direction opposite to the magnetic field strength, which *raises* the energy of the hydrogen nucleus. Electrons, neutrons and several other atomic nuclei ($^{19}_{9}F$, $^{32}_{15}P$) also have a spin quantum number equal to $\frac{1}{2}$ and behave like protons. All such particles have two possible orientations in a magnetic field, corresponding to two different energy levels.

The nuclei of deuterium ($^{2}_{1}H$) and nitrogen ($^{14}_{7}N$) have a nuclear spin $I = 1$, with *three* possible orientations in a magnetic field. These orientations are distinguished as $m_I = +I = +1$ (decrease of energy), $m_I = I-1 = 0$ (energy unaltered), and $m_I = -I = -1$ (increase of energy).

The general rule is that an atomic nucleus with a nuclear spin I has $2I+1$ possible orientations in a magnetic field corresponding to $2I+1$ different energy levels distinguished as $m_I = +I, I-1, I-2, \ldots 2-I, 1-I, -I$. For the lithium nuclei in Fig. 10.1, with $I = \frac{3}{2}$, there are thus four possible energy levels: $m_I = \frac{3}{2}, \frac{1}{2}, -\frac{1}{2}$, and $-\frac{3}{2}$.

The energy of a level designated by m_I is given as

$$E = -g m_I \mu_N H \qquad (10.1)$$

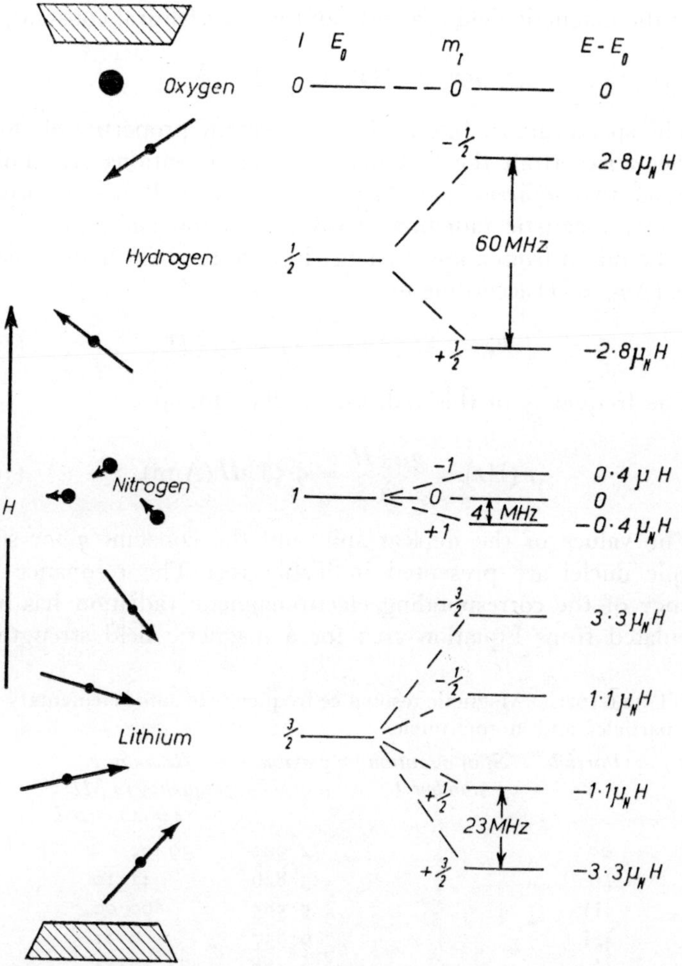

FIG. 10.1. Nuclear spins in a magnetic field. H = magnetic field strength. I = nuclear spin quantum number. E_0 = nuclear energy in the absence of a magnetic field. m_I = designation of energy levels. E = nuclear energy in a magnetic field. μ_N = constant. The direction of the arrows indicates the orientation of the nuclear magnets in the magnetic field. The length of the arrows is a measure of the strength of the nuclear magnets and corresponds to the separation between adjacent energy levels. The frequency of the resonance transition has also been indicated.

where g is a constant determined by the atomic nucleus in question, H is the magnetic field strength, and μ_N is a natural constant,

$$\mu_N = 6{\cdot}346 . 10^{-33} \text{ J m/A}$$

The spectrometric interest in the magnetic properties of atomic nuclei arises from the fact that spectral transitions are allowed between two *adjacent* energy levels in Fig. 10.1. By the absorption of electromagnetic radiation of correct photon energy a nucleus may be raised from a lower energy level to the next higher energy level ($\Delta m_I = 1$) according to

$$h\nu = \Delta E(\Delta m_I = 1) = g\mu_N H \qquad (10.2)$$

The frequency of this radiation is then found as

$$\nu\,(\text{Hz}) = \frac{g\mu_N H}{h} = 9{\cdot}58\, gH\,(\text{A/m}) \qquad (10.3)$$

The values of the nuclear spin and the constant g for some atomic nuclei are presented in Table 10.1 The resonance frequency of the corresponding electromagnetic radiation has been calculated from Equation 10.3 for a magnetic field strength of

TABLE 10.1. Magnetic resonance frequency of some elementary particles and atomic nuclei.

Particle	*Spin quantum number I*	*g-value*	*Resonance frequency in MHz at 14092 oersted*
e^-	$\frac{1}{2}$	2·002	39492
${}^{1}_{0}n$	$\frac{1}{2}$	−3·826	41·10
${}^{1}_{1}H$	$\frac{1}{2}$	5·585	60·00
${}^{2}_{1}H$	1	0·857	9·21
${}^{7}_{3}Li$	$\frac{3}{2}$	2·170	23·31
${}^{10}_{5}B$	3	0·600	6·45
${}^{12}_{6}C$	0	—	—
${}^{13}_{6}C$	$\frac{1}{2}$	1·404	15·08
${}^{14}_{7}N$	1	0·404	4·33
${}^{16}_{8}O$	0	—	—
${}^{19}_{9}F$	$\frac{1}{2}$	5·255	56·45
${}^{32}_{15}P$	$\frac{1}{2}$	2·261	24·29
${}^{127}_{53}I$	$\frac{5}{2}$	1·117	12·00

$1{\cdot}12 \cdot 10^6$ A/m. (According to Appendix II this is equal to a magnetic field strength of 14 092 oersted, which corresponds to a magnetic *induction in air* of 14 092 gauss; this value is frequently quoted in the literature.) The frequency of *nuclear magnetic resonance* (NMR) radiation varies from 4·34 MHz for nitrogen to 60 MHz for protons. Consequently, the wavelength varies from 60 to 5 metres (radiowaves).

10.2. Intensity and line width in NMR spectrometry

Two important parameters of any spectrometric method of analysis are the intensity and the width of the spectral transitions, which determine the sensitivity of the method and the resolution attainable. A discussion of these parameters for NMR spectrometry will reveal some unexpected aspects with important practical consequences.

As will be seen in the next section NMR spectra are measured in absorption, and we recall from § 1.5 that the strength of an absorption transition is determined by the number of particles in the ground state. For molecular absorption spectrometry in the visible and infra-red region (Chapters 4 and 5) this number is virtually equal to the total number of particles present; in other words, the competing process of stimulated emission (§ 1.5) may be ignored. In NMR spectrometry the situation is markedly different.

Equation 10.1 tells us that in the absence of a magnetic field all nuclear magnetic energy levels coincide, which may be visualized as the nuclear magnets being orientated completely at random, irrespective of their m_I-value. When the atoms are subjected to a magnetic field this '*degeneracy*' of the energy levels is removed: nuclei of higher m_I-values point their magnets in the direction of the magnetic field and decrease their energy, whereas nuclei of lower m_I increase their energy. After an interval (called the *relaxation time*), an equilibrium results, with more nuclei in the lowest energy level than there are in the higher energy levels, in accordance with Boltzmann's law (§ 1.5). If n_0 is the number of nuclei in the ground state and n_1 is the number of nuclei in the next higher energy level, Boltzmann's equation (Equation 1.6), combined with Equation 10.2, gives

$$\frac{n_1}{n_0} = \exp\frac{-\Delta E}{kT} = \exp\frac{-g\mu_N H}{kT} \qquad (10.4)$$

At room temperature (T = 300°K) this yields

$$\frac{n_1}{n_0} = \exp(-1{\cdot}5\,.\,10^{-12}\,gH) = 1 - 1{\cdot}5\,.\,10^{-12}\,gH$$

The overpopulation of the ground state with respect to the next higher energy level is then given by

$$\frac{n_0}{n_1} - 1 = 1{\cdot}5\,.\,10^{-12}\,gH \qquad (10.5)$$

Substituting a magnetic field strength of the order of 10^6 A/m and a *g*-value of about one, we find that for every million nuclei there is only one nucleus more in the ground state than there is in the first excited state. Now, if the system is irradiated with electromagnetic radiation of resonance frequency, only one out of each million atomic nuclei will contribute to the measured absorption signal! The absorption of the other nuclei in the ground state is completely cancelled by stimulated emission from the nuclei in the excited state (§ 1.5). Therefore, the intensity of the absorption spectrum is comparatively small, and nuclear magnetic resonance is a rather insensitive method of spectrometric analysis.

According to Equation 10.5 the overpopulation of the ground state and hence the absorption intensity increases linearly with the magnetic field strength. This explains why over the years the magnetic field strength employed has increased steadily from 14000 to 23000 and, recently, to 51000 oersted (1·12, 1·87 and 7·1 . 10^6 A/m, corresponding to proton resonance frequencies of 60, 100 and 220 MHz, respectively).

Equation 10.5 also shows that at fixed magnetic field strength the absorption intensity increases with the value of g or with increasing frequency of the resonance radiation (Equation 10.3). Fig. 10.2 shows schematically the absorption of equal numbers of different nuclei at a constant field strength of 1·12 . 10^6 A/m. The resonance

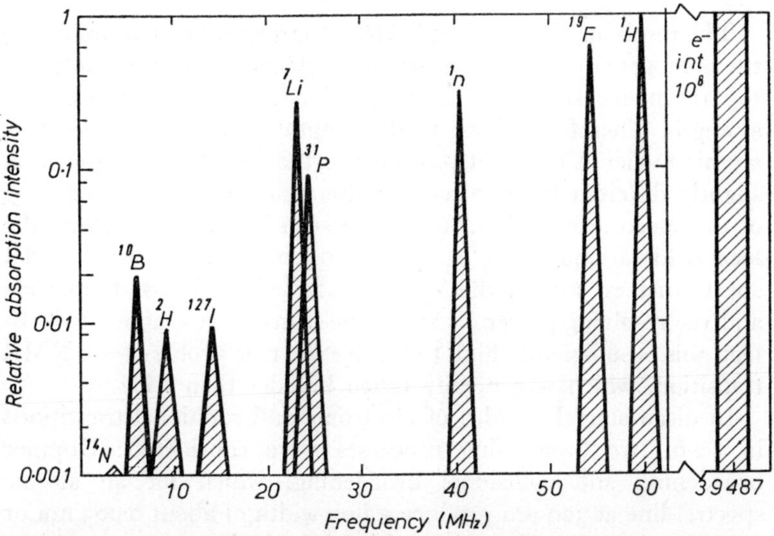

FIG. 10.2. Nuclear magnetic resonance frequencies for a few atomic nuclei in a magnetic field of 14000 oersted ($1{\cdot}12 \cdot 10^6$ A/m).

frequencies are taken from Table 10.1. Indeed, the relative strength of the absorption increases appreciably with the frequency, although this is apparently not the only factor (compare Equation 10.6). For elements such as boron or nitrogen about 10^{21} atomic nuclei per gram sample are required for a good NMR signal which means that these elements must be present to the extent of several per cent in the sample. Accordingly, the *inorganic* applications of nuclear magnetic resonance spectrometry are limited to theoretical studies of pure, usually solid, substances.

The important *analytical* applications of nuclear magnetic resonance spectrometry deal with organic substances. The most abundant isotopes of the elements oxygen and carbon have no nuclear spin (the less abundant isotope $^{13}_{6}C$ with nuclear spin $\frac{1}{2}$ is currently becoming of interest, however), and for nitrogen the signals are generally too weak, because of the small g-value. However, nearly all organic compounds contain hydrogen, which has the advantage of a very large g-value and a correspondingly strong absorption. Consequently, proton resonance can be observed in solution. The following sections will deal predominantly with proton resonance spectrometry.

The resolving power of an NMR spectrometer is determined by the homogeneity of the magnetic field. According to Equation 10.3 the resonance frequency is proportional to the magnetic field strength. Therefore, if the field strength varies over the sample, atomic nuclei at different positions in the sample will resonate at slightly different frequencies and this will produce a broadening of the spectral transition. As will be seen in the next section, the local field variations can be reduced to 0·01 A/m in modern NMR spectrometers, so that the 'spectral bandwidth' is less than 1 Hz and the resolving power ($\nu/\Delta\nu$) of the instrument is about 10^8! In fact, this is sufficiently high to resolve the true profile of the NMR transition, which is generally much broader than 1 Hz.

In discussing the width of electronic and rotational transitions in § 2.6, two broadening processes were considered: Doppler broadening and collisional broadening, which for an atomic spectral line at 300 nm produce a line width of about 0·005 nm or 10^{10} Hz. Actually, there is a third broadening process, which, acting in isolation, would make the width of a spectral transition (in frequency units) equal to the reciprocal of the relaxation time. For atomic spectral lines the relaxation time is the reciprocal of the transition probability, which is of the order of $10^8\ s^{-1}$ (§ 6.3). Accordingly, the *natural width* of atomic spectral lines is about 10^8 Hz, which is completely negligible compared to Doppler broadening or collisional broadening.

In NMR spectrometry, however, the width of the spectral transition is determined purely by relaxation broadening. In solids, the relaxation time is about 10^{-4} s and the line width is about 10^4 Hz, which is much larger than the spectral bandwidth of the NMR spectrometer. In fact, the recorded line profiles offer an opportunity to study the relaxation processes in solids. In solution, however, the relaxation time for protons can reach values of the order of one second, and thus the line width is reduced to 1 Hz. It is, in fact, this small line width which makes NMR spectrometry an important tool for chemical analysis, because it allows the observation of fine structure in proton resonance spectra; the single spectral transition of the proton drawn in Fig. 10.2 is now split into a number of separate sharp lines, as shown in Fig. 10.4.

The large relaxation time of proton resonance in liquids has another consequence. Through absorption of electromagnetic radiation, protons are excited from the ground state to the higher energy level. This reduces the small overpopulation of the ground state even further. Because of the large relaxation time the equilibrium situation is not rapidly restored. Therefore, if sample irradiation at fixed frequency is prolonged, the absorption signal is seen to decrease. Ultimately, the population of the two energy levels becomes equal, and no (net) absorption can be measured. This is called *saturation*.

10.3. The nuclear magnetic resonance spectrometer

The basic design of an instrument for recording nuclear magnetic resonance spectra is shown in Fig. 10.3. The sample is dissolved in carbon tetrachloride or deuterated chloroform (C^2HCl_3), and about 0·5 ml in a small glass tube is placed between the poles of a strong magnet. At both sides of the tube, and perpendicular to the magnetic field, two coils generate a radiowave of, for example, 60 MHz. A second coil around the tube serves to detect the radiowaves. In this *crossed-coil* spectrometer the transmitting and receiving coils are mutually perpendicular. If the sample does not absorb radiation, the receiving coil does not detect any signal. However, if the sample does absorb radiation, the receiving coil detects a signal proportional to the amount of radiation absorbed. In the alternative arrangement, the single-coil spectrometer, the same coil is used both for transmitting and receiving the radiowaves.

As was discussed in the previous section, the magnetic field must meet very strict requirements. The magnetic field strength should be high, to produce strong absorption signals, but it must also be uniform over the entire sample in order that all protons resonate at the same frequency and a small line width is produced. Strong and homogeneous magnetic fields of 10^6–10^7 A/m (10000–50000 oersted) can be realized in practice both with permanent magnets and with electromagnets. Local variation of the magnetic field over the sample is limited by secondary coils (for electromagnets) to about 1 A/m (a few milli-oersted). The effective homogeneity of the field is further increased by rapid rotation of the sample tube

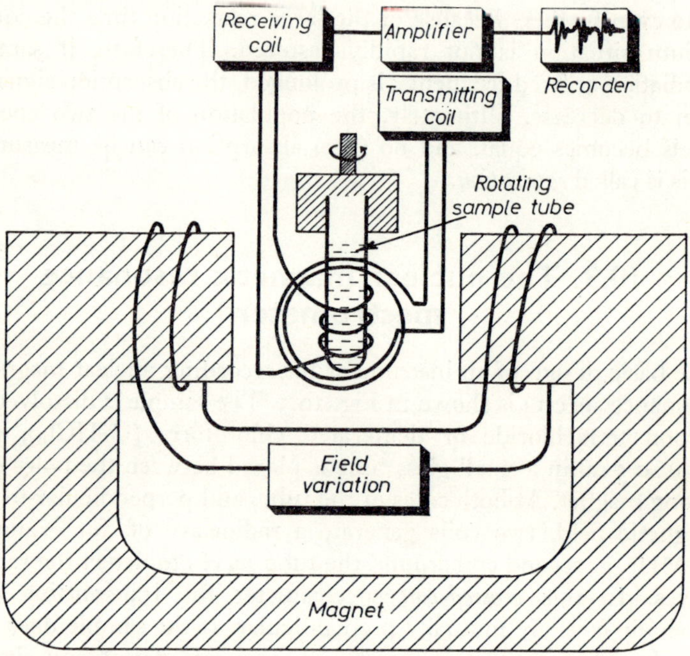

FIG. 10.3. Schematic diagram of a nuclear magnetic resonance spectrometer.

(for example at 1500 revolutions per minute). The local field variations are thus reduced to 0·01 A/m (10^{-4} oersted).

The magnetic resonance spectrum may be recorded in two ways. Either the magnetic field strength can be varied at a fixed radiofrequency, or the frequency can be varied at a fixed magnetic field strength. For studying the fine structure of proton resonance spectra the total range of the scan (the 'sweep') is only some 500 Hz at a resonance frequency of 60 MHz (or 10 A/m, around a basic field strength of 10^6 A/m). For other nuclei the sweep can be several kHz. By either method the nuclear magnetic resonance spectrum is obtained as a series of peaks on a recorder chart.

The deflections of the recorder pen are a measure of the amount of radiation *absorbed* by the sample. Since the absorption is very weak, Beer's law yields

$$I_a = I_0 - I_t = I_0(1-10^{-abc}) = 2{\cdot}3\, I_0 abc \qquad (10.6)$$

(compare § 4.8).

The deflection of the recorder pen is proportional to the concentration c of absorbing (resonating) nuclei.

In order to avoid the effect of *saturation* due to prolonged irradiation (§ 10.2), the spectrum must be scanned fairly rapidly. A typical scan rate for proton spectra is 2 Hz per second. As the line width is about 1 Hz, each line is recorded in 0·5 seconds. Thus the recorder pen must show a rapid response, and damping must be minimal. As a result, noise is rather large, and weak absorption signals are not easily distinguished from background fluctuations.

Because the noise cannot be suppressed by damping (since this requires lower scan rates, leading to saturation), another solution is used in recording *weak proton resonance spectra*, where the output of the spectrometer is fed into a computer. The spectrum is scanned rapidly and without damping, and the information is stored in the computer; next, the spectrum is scanned for a second time and the absorptions are added to the previous values by the computer. This is repeated over and over again. The random noise fluctuations are smoothed out, but the absorption peaks are enhanced. More exactly, for N repeated scans, true absorption lines increase with N, whereas the noise increases only with $\sqrt{N}$. After the preset number of scans has been run, the summed total spectrum is supplied to the recorder.

Originally, this technique of computer-averaging or *ensemble-averaging* was only applied in nuclear magnetic resonance and far infra-red spectrometry, but obviously the technique is quite generally applicable to other fields of spectrometry in which weak signals must be measured (for example flame spectrometry). On the other hand, the technique should not be used indiscriminately. Computers are still fairly expensive. Moreover, since the effective enhancement of the signal is proportional to $\sqrt{N}$, a rather large number of scans is required significantly to improve the results, and this may take a long time. For example, a tenfold improvement of a spectrum scanned in 10 minutes takes 100 runs, or 16 hours. Of course, this may be done overnight, but still requires that the instrument remains stable for so long a period of time.

In nuclear magnetic resonance this means that the magnetic field strength must remain constant down to the 0·2 milli-oersted required for small line width. This cannot be realized in practice, but it is possible to control the magnetic field strength automatically, by *locking* the field to the constant resonance frequency of an added reference compound. Tetramethylsilane, $Si(CH_3)_4$ (TMS), is widely used, as it shows a single sharp resonance line (Fig. 10.4b).

10.4. Chemical shift in proton resonance spectra

It would appear from Equation 10.3 that all protons resonate at one and the same frequency. Indeed, with a spectrometer of low resolving power all organic substances are observed to absorb at the same frequency. This is shown in Fig. 10.4a by the single broad band obtained for a mixture of tetramethylsilane and ethanol dissolved in deuterated chloroform (C^2HCl_3).

However, if the resolving power of the instrument is improved and the width of the absorption transition decreases, it appears that the absorption lines of different substances no longer coincide. Even a pure compound may give rise to a number of well separated absorption lines, as is shown by the example of ethanol in Fig. 10.4b.

In order to explain this phenomenon, it must be recalled that Equation 10.3 has been derived for an isolated proton experiencing the direct influence of the external magnetic field. In reality, of course, the hydrogen nuclei in a molecule are surrounded by electrons moving in molecular orbits around the hydrogen nucleus and adjacent nuclei. Under the influence of the external magnetic field the electronic orbits are slightly altered, and this induces a small magnetic field of *opposite sign*. As a result the magnetic field acting upon the proton is slightly smaller than the external magnetic field H_0:

$$H = H_0(1 - \sigma) \qquad (10.7)$$

where σ is called the *shielding factor*, to indicate that the electrons shield the nucleus from the external field.

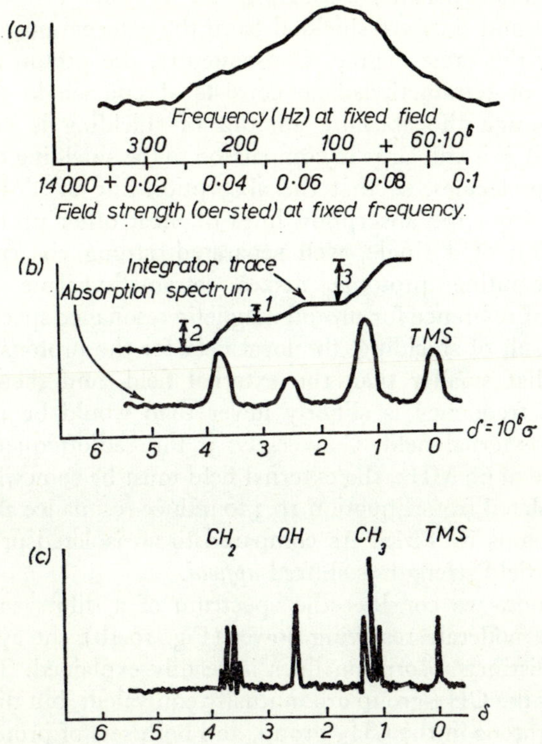

FIG. 10.4. Proton resonance spectrum of ethanol. (a) Very low resolving power. (b) Moderate resolving power; chemical shift. (c) High resolving power; spin–spin coupling. The frequency scale in the top drawing runs from $60 . 10^6$ to $60 . 10^6 + 300$ Hz; the field strength scale runs from 14000 to 14000·1 oersted. In the bottom two drawings the absorption spectrum is plotted as a function of the chemical shift parameter δ, explained in the text.

As might be expected, the amount of shielding depends upon the local electron density around the nucleus, and this is determined by the chemical environment of the nucleus. Accordingly, the variations in absorption frequency between protons of different molecules, or between protons of different groups in one molecule, are known as the *chemical shift*.

In tetramethylsilane, $Si(CH_3)_4$, all protons are chemically equivalent and thus are shielded from the external magnetic field to exactly the same degree. Consequently, the proton resonance spectrum of tetramethylsilane consists of one single absorption line. Although the absolute amount of shielding is very small ($\sigma \simeq 10^{-5}$), it is still large in comparison to the shielding of protons in other molecules, so that the absorption line of TMS is well separated from the absorption lines of most other protons. The combination of a single, well separated, strong absorption line (12 contributing protons) makes tetramethylsilane an ideal standard of reference for proton magnetic resonance spectrometry.

As a result of shielding, the local field for the protons in TMS is somewhat smaller than the external field, and therefore the resonance frequency is slightly lower than would be calculated from the external field. Conversely, if the radiofrequency has a fixed value of 60 MHz, the external field must be somewhat larger than calculated from Equation 10.3 to induce resonance absorption of the protons in TMS. As compared to an isolated proton, the resonance field strength is shifted *upfield*.

When now we consider the spectrum of a dilute solution of ethanol at moderate resolving power (Fig. 10.4b), the appearance of three distinct absorption lines is readily explained. The three protons of the CH_3-group are mutually equivalent, but differ from the two protons in the CH_2-group, and both sets of protons differ chemically from the single proton of the OH-group. Obviously, these three types of protons have different shielding factors, and, at a higher resolving power, the broad absorption band of ethanol in Fig. 10.4a is resolved into three separate peaks.

An important secondary observation is that the areas of the three absorption peaks in the spectrum of ethanol are proportional to the number of equivalent protons contributing to the peak; that is, they are in the ratio 3 : 2 : 1. This is due to the fact that the absorption coefficient in Equation 10.6 is equal for all protons. Therefore, an integrator trace, simultaneously recording the peak areas of the absorption spectrum, forms an important aid in deducing the molecular structure from the nuclear magnetic resonance spectrum. It is readily seen from Fig. 10.4b that the peak with the lowest shielding (*downfield* shift at fixed radiofrequency) is due to CH_2-protons, and the peak with the highest shielding is due to CH_3-protons. The peak from the OH-proton lies at an intermediate field strength (see however Fig. 10.6).

The total range of the chemical shift is very small. At a fixed field strength of 14000 oersted the resonance frequencies of all organic protons fall within a range of 1000 Hz, and at a fixed frequency of 60 MHz the necessary magnetic field sweep is only 0·2 oersted. According to Equation 10.7 the shift is equal to $\Delta H = H_0 - H = \sigma H_0$, and thus is proportional to the external magnetic field strength. Therefore, increased field strengths offer an extended spectral range and better resolution. This is the main reason why proton resonance spectrometers operating at 23000 oersted (100 MHz) or even 51000 oersted (220 MHz) are coming into use.

On the other hand, it is not the absolute magnitude of the chemical shift (in frequency or in field strength units), but the shielding factor σ that is significant for the structural position of a proton in a molecule. Because it is difficult to measure the shielding factor directly, the chemical shift is measured relative to a reference substance, usually TMS, which is added in small proportions to the sample solution. The chemical shift is now defined as

$$\delta = 10^6(\sigma_{TMS} - \sigma) \qquad (10.8)$$

The δ-value of TMS is by definition equal to zero, while in Fig. 10.4b ethanol shows proton resonance peaks at δ-values of 1·2 (CH_3), 2·6 (OH) and 3·7 (CH_2). Note that a high value for δ indicates a *downfield* shift, whereas a low value for δ represents an *upfield* shift.

10.5. Spin–spin coupling

The width of the absorption peaks in Fig. 10.4b is about 15 Hz. If the resolving power is increased still further, the absorption peaks of ethanol are split into more complicated patterns. The absorption peak of the CH_3-protons is split into three lines, and the peak of the CH_2-protons is split into a quartet of four lines (Fig. 10.4c). This is caused by *spin–spin coupling*.

The protons in the CH_2-group behave like tiny bar magnets, and produce a magnetic field, which is very much weaker than the external field, but still exerts an influence upon the local magnetic field experienced by the CH_3-protons. Now, if the proton-magnets of the CH_2-group are in opposite directions (opposite spins) their effect is nullified and their influence upon the field at the CH_3-group is nil. This can clearly happen in two ways (Fig. 10.5).

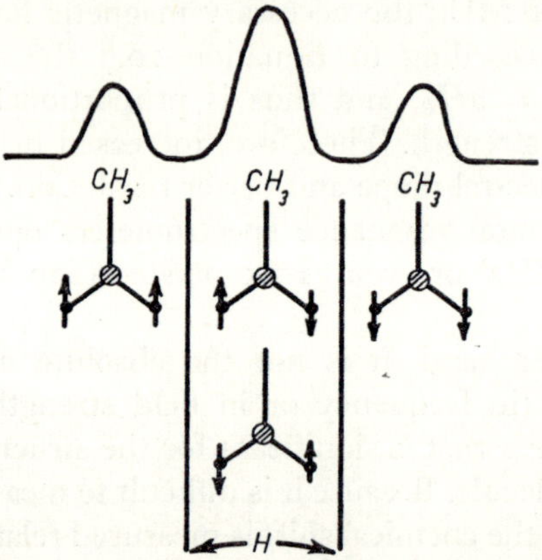

FIG. 10.5. Triplet-splitting of the CH_3-absorption under the influence of the spin orientations of the CH_2-protons.

However, if the spins of the CH_2-protons are parallel, the magnetic field at the CH_3-protons is either enhanced or reduced. In a large number of ethanol molecules these three possibilities will occur in the ratios 1 : 2 : 1, and therefore the absorption peak of the CH_3-protons is split into a triplet with intensity ratios 1 : 2 : 1.

In an exactly similar way the spins of the CH_3-protons influence the field at the CH_2-group. Now there are three proton-magnets and it is readily seen that the CH_2-peak will be split up into four lines with intensity ratios 1 : 3 : 3 : 1. In general, a group of n protons produces a splitting of the resonance peak of neighbouring protons into $n+1$ components. Note that the occurrence of spin–spin coupling is not determined by the relative position of the absorption peaks in the spectrum, but by the relative position of the protons in the molecule. In saturated compounds only those spins couple which belong to adjacent proton groups. In unsaturated compounds the coupling may extend over several atomic bonds. If a proton group is coupled to n protons on the one side and to m protons on the other side, the theoretical number of multiplet is $(n+1)(m+1)$. In practice the number of observed lines is frequently smaller, because lines may coincide.

So far the explanation of spin–spin coupling effects is straightforward, but now we run into two problems. In the first place we might wonder why no multiplet splitting is observed for the absorption peak of TMS although the different methyl groups in this molecule are adjacent. The reason is that protons with the same chemical shift δ do not couple mutually. Therefore, the twelve protons in TMS, the six protons in benzene, the three protons in methyl bromide and the six protons in ethane all give rise to a single sharp spectral line. On the other hand, all three protons in $CH_2{=}CHCl$ show different chemical shifts, because each proton is at a different distance from the chlorine atom. Also, each proton is coupled to the other two, so that we might expect a proton resonance spectrum of three quartets.

A second problem is the single absorption line observed for the OH-proton in Fig. 10.4c. Because the OH-proton is next to the CH_2-group, we would expect the absorption peak of the OH-proton to be split into a triplet. Moreover, we would expect the absorption peak of the CH_2-protons to be split into eight lines by simultaneous coupling to one proton on one side and to the three methyl protons on the other side. This is indeed observed, if the ethanol is carefully dried, as is shown by the spectrum in Fig. 10.6a, where it may also be noted that the OH-absorption is shifted downfield to larger δ. However, in the presence of traces of water no coupling of the OH-proton is observed.

This is explained as follows. In the presence of water the hydroxyl proton moves very rapidly from one molecule to the next through the formation of hydrogen bridges. Consequently, during the relaxation time of one second, the OH-proton finds itself attached to many different ethanol molecules, and it 'sees' all possible spin positions of the CH_2-protons. As a result, the OH-proton experiences an 'average' spin effect equal to zero, and the absorption peak is not split. In turn, this rapid exchange excludes a possible influence of the spin of the hydroxyl proton upon the CH_2-absorption peak.

A similar phenomenon explains the single absorption line observed for acetyl chloride (CH_3–COCl). If the molecule were rigid at least one of the three methyl protons would be at a different

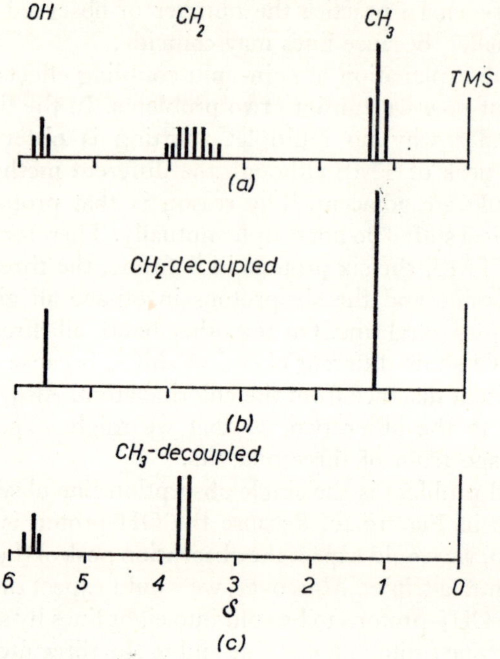

FIG. 10.6. Spin-decoupling in the spectrum of ethanol. (a) Spectrum of carefully dried ethanol. (b) CH_2-spins decoupled. (c) CH_3-spins decoupled.

distance from the chlorine and oxygen atoms. In reality, there is a rapid free rotation around the C–C axis, and all three protons experience an average shielding effect, so that they have the same chemical shift and, of course, no spin–spin coupling.

The rate of these positional changes frequently depends upon the temperature and, therefore, the structure of a proton resonance spectrum may vary with the temperature of the solution. In turn, nuclear magnetic resonance spectrometry is a valuable tool for studying the kinetics of such processes.

Let us return now to the spectrum of ethanol. The mutual separation of the components in the triplet and in the quartet of Fig. 10.4c is 7 Hz. This is expressed by saying that the spin–spin coupling constant J between the methyl group and the methylene group is 7 Hz. This is much smaller than the difference in chemical shift, which in Fig. 10.4 amounts to 150 Hz for the CH_3 and CH_2 protons. Indeed, the simple rules for multiplet splitting given above are only valid if the difference in chemical shift between two proton groups is some twenty times larger than their spin–spin coupling constant. In that case the proton resonance spectrum is an orderly array of well separated multiplets.

However, if the ratio of the difference in chemical shift to spin–spin coupling constant is less than 20, the above rules for multiplet splitting are no longer obeyed. Although the total number of lines remains the same, neither their relative position nor their relative intensity can be easily predicted. For some simple cases other splitting schemes can be derived, but more complicated cases require extensive quantum mechanical calculations. From a practical point of view, the similarity between spin–spin coupling constants and differences in chemical shift frequently gives rise to a complex proton resonance spectrum. A substantial amount of experience is then required to discover the structure of the spectrum (and of the sample compound !).

Fortunately, there are two practical tools for reducing the confusion between spin–spin coupling and chemical shift. In contrast to chemical shift, spin–spin coupling constants do *not* depend upon the external magnetic field strength. Therefore, increasing the magnetic field strength also increases the ratio of differences in chemical shifts to spin–spin coupling constants, which simplifies the spectrum. Also, if the spectrum is recorded at two different field strengths (for example 14 000 and 23 000 oersted), two peaks with different chemical shifts will be further separated in the stronger field, whereas the separation of multiplet lines remains unaltered. Note, however, that this requires two different (and differently priced !) instruments.

Perhaps more important is the possibility of *decoupling* spin–spin interactions. Fig. 10.6a presents the proton resonance spectrum of dry ethanol which as explained above consists of two triplets and an octet. Now, with a second transmitting coil a strong beam of radiowaves can be generated with the resonance frequency of the CH_2-protons. The proton resonance of this group becomes saturated and the methylene protons very rapidly change their spin position by the alternative processes of absorption and stimulated emission. In the same way as described above for the exchange of the OH-proton this removes the interaction of the CH_2-protons with the other protons in the molecule. Because the OH-proton and the CH_3-protons are too far apart for coupling, the spectrum is reduced to only two lines with an intensity ratio of 1 : 3 (Fig. 10.6b). In the same way decoupling of CH_3-protons reduces the spectrum to a doublet and a triplet (Fig. 10.6c).

10.6. Applications to organic analysis

Typically, 0·5 ml of a 10 per cent sample solution are used in recording the proton resonance spectrum of an organic compound. This amounts to about 50 mg of sample. Special techniques, such as computer averaging of spectra, permit a reduction of the sample size to 0·5 mg. Still, it is clear that nuclear magnetic resonance spectrometry is appreciably less sensitive than infra-red spectrometry (Chapter 5) or mass spectrometry (Chapter 11).

Proton resonance spectrometry is a most valuable tool in determining the structure of unknown and newly-synthesized compounds. The proton resonance spectrum frequently permits a rapid choice between a number of possible molecular structures. The following spectral data are taken into consideration.

(*a*) The chemical shift of a peak indicates the nature of a proton group (aromatic protons, CH_3-group, unsaturated carbon groups, etc.). A few examples are presented in Table 10.2 (p. 232). In addition to the group itself, neighbouring atoms (notably oxygen) exert an influence, and chemical shifts may give an indirect indication of the presence of functional groups in the molecule.

(*b*) The multiplet structure of a peak is determined by spin–spin coupling with adjacent protons. Consequently, multiplet patterns provide information about the connection between different proton groups in the molecule. Values of coupling constants between various proton groups have been tabulated and can be found in textbooks. Spin decoupling is applied in unravelling complex spectra.

(*c*) The integrator trace in the spectrum indicates the number of equivalent protons involved in an absorption peak. This trace allows a count of the number of protons responsible for a chemical shift.

As a simple illustration of the application of proton resonance spectrometry to structure determinations, we consider the spectrum in Fig. 10.7. The integrator trace shows that the four peaks in the spectrum have relative intensities of 1 : 2 : 2 : 5. It is very probable that these numbers also indicate the number of protons

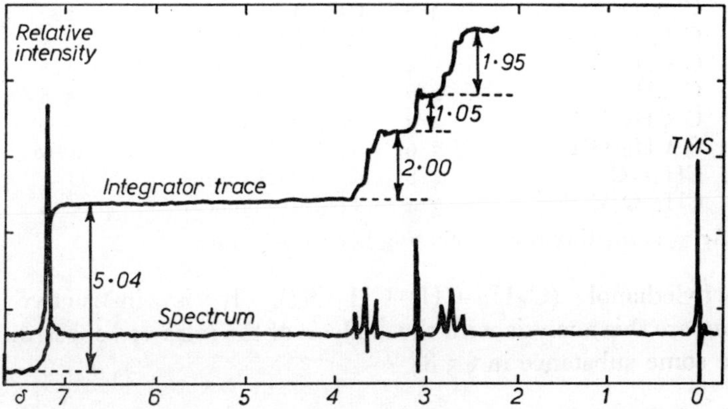

FIG. 10.7. Proton resonance spectrum of an unknown compound; explanation in text. (The kindness of Drs. A. van Veen, Laboratory for Organic Chemistry, Delft, in providing this spectrum is gratefully acknowledged.)

involved in the respective peaks. From the chemical shift of the largest peak we conclude that this peak is due to aromatic protons, and because five protons are responsible for it, this indicates a monosubstituted benzene ring. Two protons are involved in each triplet and, consequently, these are due to CH_2-peaks. The triplet splitting (approximately 1 : 2 : 1) indicates that these groups are only coupled to each other and must be located next to each other in the molecule. This is confirmed by the coupling constant of 7 Hz. It follows from the chemical shift that the CH_2-group at $\delta = 2{\cdot}7$ is connected to the aromatic ring (these two groups do not couple), but the other CH_2-resonance is shifted very far downfield. According to Table 10.2 this may be due to the vicinity of an oxygen atom. Because the last *single* peak in the spectrum is due to one proton only, we suspect the presence of an OH-group. Indeed, the OH-resonance of a concentrated solution may very well be located at $\delta = 3{\cdot}1$. The unknown compound is thus found to be

TABLE 10.2. Chemical shift of proton groups.

Proton group	δ	*Proton group*	δ
CH_3–C–	0·9	C–CH–C	1·5
CH_3–C–O	1·4	CH–Ar	3·0
CH_3–CO–C	2·2	CH–OH	3·9
CH_3–Ar	2·3	CH=C	4·5–6
CH_3–O–Ar	3·8	CH≡C	2–3
CH_3–C=C–	1·6	R–OH (diluted)	0·5
C–CH_2–C	1·4	R–OH (concentrated)	3–5
C–CH_2–Ar	2·7	Ar–OH	5–8
C–CH_2–CO–C	2·4	Ar–H	6–9
C–CH_2–OH	3·6	Ar–CHO	9–10
CH_2=C	4·8	R–COOH (dimer)	11–12
CH_2–O–C	3·4	TMS (reference)	0

Ar = aromating rings including heterocyclic ones.

phenylethanol (C_6H_5–CH_2–CH_2OH). It is instructive to compare this analysis with the analysis of the infra-red spectrum of the some substance in § 5.8.

10.7. Electron paramagnetic resonance

As has been remarked in § 10.1, electrons also have a spin quantum number $\frac{1}{2}$. Accordingly, a single electron subjected to a strong magnetic field will have two possible magnetic orientations and two different energy levels at its disposal (Fig. 10.1). Again, the spectral transition between these levels is allowed, and the frequency of the resonance radiation is given by

$$\nu = \frac{g\mu_e H}{h} \qquad (10.9)$$

(compare Equation 10.3). The g-value of an electron is equal to 2 (Table 10.1), but the constant μ_e is about 2000 times larger than μ_N as a result of the smaller electronic mass. In the commonly applied magnetic field of $0{\cdot}26 \cdot 10^6$ A/m (3300 oersted) the resonance frequency of the electronic spin transition is thus found to be 9400 MHz. The corresponding wavelength is about 3 cm, so the *electron paramagnetic resonance* (EPR) spectrometer utilizes the microwaves familiar, for example, in radar techniques. In consequence of the comparatively large value of μ_e the overpopulation of the ground state is much larger than in nuclear magnetic resonance (§ 10.2) and hence the absorption signal is very much stronger (Fig. 10.2). A concentration of only 10^{12} free electrons per gram sample can still be detected.

Now, of course, we have no free electrons but only bound electrons in atoms or molecules, and before we can evaluate the potential applications of EPR spectrometry we must have a closer look at these complex systems. The electrons in an atom describe orbits around the nucleus; each orbit is characterized by three quantum numbers (§ 2.2), and can accommodate two electrons. If there is only one electron in the orbit (as is the case for the hydrogen atom or for the valence electron in alkali atoms), the spin of the electron may have either of two values ($s = +\frac{1}{2}$ or $s = -\frac{1}{2}$) and in a magnetic field two energy levels result. However, if the orbit is occupied by two electrons, these electrons must be distinguishable according to Pauli's exclusion principle (§ 2.2); therefore their spins must be opposite, and must remain opposite. According to quantum mechanics the total electron spin of this orbit is zero. Consequently, if all electrons in an atom are *paired*, the total electron spin of the atom is zero and its energy is unaffected by the presence of a magnetic field (except, of course, for possible nuclear spin effects).

In a molecule each molecular orbit is always occupied by two paired electrons of opposite spin. In fact, the valence sign in elementary chemical structure formulae represents precisely such a pair of electrons. Therefore, the total electron spin of a *molecule in the ground state* is zero and, again, its energy is unaffected by a magnetic field. Consequently, ground state molecules, as well as atoms with paired electrons only, cannot be studied by electron paramagnetic resonance spectrometry.

Obviously then, the presence of one or more unpaired electrons is an essential requirement for the application of EPR spectrometry. This occurs in free hydrogen and alkali atoms at elevated temperatures (for example in flames), but it occurs also in certain crystal lattices, where transition elements, such as chromium or manganese, can have up to five unpaired electrons with parallel spins (obviously, these electrons move in different orbits). EPR spectrometry is often applied to study these elements.

A molecule with one unpaired electron is called a radical. Such a radical will have a resultant electron spin of $\frac{1}{2}$, so that it can be observed with high sensitivity by EPR spectrometry. An obvious application is the study of chemical reactions, which frequently involve radicals as intermediary products. The fine structure of the EPR spectrum that results from the interaction of the electron spin with nuclear spins provides information about the structure of the radical, and the location of the unpaired electron in it.

As was remarked above, a molecule in the ground state has a total electron spin of zero, because all electrons are paired. However, the molecule may be excited to a higher energy level where two electrons describe different orbits. Now their spins may be either opposite or parallel, and in the latter case the total spin quantum number of the molecule is $\frac{1}{2}+\frac{1}{2} = 1$. In the same way as with the nuclear spin of nitrogen (Fig. 10.1) this gives rise to three energy levels in a magnetic field, with allowed spectral transitions between two adjacent levels. Electron paramagnetic resonance spectrometry thus permits the investigation of the so-called triplet levels of molecules.

Bibliography

1. FLUCK, E., *Die Kernmagnetische Resonanz und ihre Anwendung in der anorganischen Chemie*, Springer, Berlin, 1963.
2. CHAPMAN, D., and MAGNUS, P. D., *Introduction to Practical High Resolution Nuclear Magnetic Resonance Spectroscopy*, Academic Press, London, 1966.
3. INGRAM, D. J. E., *Spectroscopy at Radio and Microwave Frequencies*, Butterworths, London, 2nd ed., 1967.
4. BIBLE, R. H., *Interpretation of NMR Spectra*, Plenum Press, New York, 1965.
5. WILLIAMS, D. H., and FLEMING, I., *Spectroscopic Methods in Organic Chemistry*, McGraw Hill, London, 1966.
6. BOVEY, F. A., *Nuclear Magnetic Resonance*, Academic Press, New York, 1969.

Problems

1. What is the number of energy levels for a molecule with three electrons in different orbits (parallel spins) in a magnetic field? What is the result if one electron recedes to an orbit occupied by one of the other two electrons?

2. Calculate the resonance frequency in a magnetic field of 14092 oersted for the following nuclei:

Nucleus	*Spin*	*g-value*
^{23}Na	$\frac{3}{2}$	1·477
^{35}Cl	$\frac{3}{2}$	0·547
^{50}V	6	0·557
^{54}Cr	?	?
^{115}In	$\frac{9}{2}$	1·224

3. Why is deuterated chloroform used as a solvent in nuclear magnetic resonance spectrometry instead of ordinary chloroform?

4. In § 10.1 it was remarked that *even–even* nuclei have zero nuclear spin. Verify in Table 10.1 that *odd–odd* nuclei have integer spins, whereas *even–odd* nuclei have half-numbered spins.

5. Explain the spectrum of Fig. 10.6c. What spectrum will result if dry ethanol is irradiated with an intense beam at the resonance frequency of the OH-proton?

6. Explain the quartet-splitting of the CH_2-peak in Fig. 10.4c under influence of the CH_3-protons.

7. Advance arguments to show that the spectrum in Fig. 10.7 cannot be due to any of the following structures:

(*a*) C_6H_5–O–C_2H_5
(*b*) C_2H_5–C_6H_4–OH
(*c*) CH_3–C_6H_4–CH_2OH
(*d*) C_6H_5–CHOH–CH_3

8. What differences would you expect between the spectra of xylene (CH_3–C_6H_4–CH_3) and ethylbenzene (C_2H_5–C_6H_5)? Discuss the number of peaks, the multiplet-splitting and the peak areas.

9. A substance with the formula C_3H_8O shows the following nuclear magnetic resonance spectrum: a doublet at $\delta = 5{\cdot}1$, a septet at $\delta = 3{\cdot}9$ and a doublet at $\delta = 1{\cdot}1$. What is its structure, and what are the ratios of the peak areas?

CHAPTER ELEVEN

MASS SPECTROMETRY

11.1. The single-focusing mass spectrometer

In the preceding chapters we have always recognized the sample components (atoms, atomic groups or molecules) by their characteristic response to electromagnetic radiation. Another highly characteristic property is the atomic mass or molecular mass. If we succeed in measuring the masses of the separated sample constituents with sufficient precision, we have at our disposal a nearly specific method of determining these constituents. An instrument that combines the required precision with a high sensitivity is the *mass spectrometer*.

The operating principle of a simple single-focusing mass

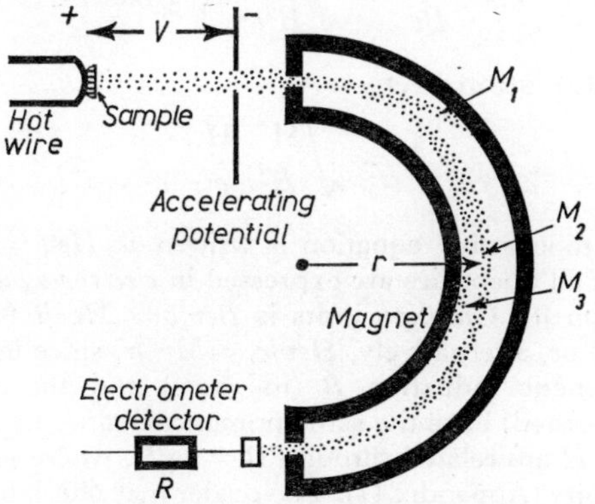

FIG. 11.1. Single-focusing mass spectrometer. The ions are generated by evaporation from a hot wire, then accelerated in an electric field, deflected by a 180° magnetic field and collected by an electrometer detector behind the exit slit; $M_1 > M_2 > M_3$. The magnetic induction is perpendicular to the drawing.

spectrometer is shown in Fig. 11.1. A small amount of an inorganic sample is pasted onto a platinum or tungsten strip, which is then heated electrically to a temperature of about 2500°K. The sample evaporates and a portion of the evaporated particles acquire a positive charge. The positive ions are accelerated by an electric potential difference of a few thousand volts, and acquire a kinetic energy of

$$\tfrac{1}{2} Mv^2 = eV \tag{11.1}$$

Here M is the mass of the ion, v is the final velocity and e is the elementary unit of charge, because the applied *thermal ionization* produces only monovalent ions.

Through an entrance slit the positive ions reach a strong magnetic field. The ions are now subjected to the Lorentz force, which acts perpendicularly to the direction of propagation and the direction of the magnetic field. Consequently, the velocity is unaffected, but the direction of propagation is changed and the ions describe a circular orbit. The radius r of this orbit follows from the equilibrium between the Lorentz force and the centrifugal force*:

$$Bev = Mv^2/r \tag{11.2}$$

whence

$$r = \frac{Mv}{Be} = \sqrt{\frac{2M}{B^2e^2}} \cdot \sqrt{\tfrac{1}{2}Mv^2} \tag{11.3}$$

and, applying Equation 11.1,

$$r = \sqrt{\frac{2V}{B^2} \cdot \frac{M}{e}} \tag{11.4}$$

* In some textbooks this equation is written as $Hev = Mv^2/r$, which is only correct if all quantities are expressed in *electromagnetic* units. The correct expression for *Gaussian* units is $Bev/c = Mv^2/r$ (where c is the velocity of light) or, alternatively, $Hev/c = Mv^2/r$, since in the Gaussian system the magnetic induction B (in gauss) and the magnetic field strength H (in oersted) have the same numerical value. In the SI system, however, B and H are related through $B = \mu_0 H$, where μ_0 is now quite different from unity (Appendix II). The reader may check for himself how these alternatives affect the following equations. It should then become clear that Equation 11.5 can be written as follows (for an electric potential of 2000 V and a radius of 25 cm):

$$M\,(\text{AMU}) = 1508\,B^2\,(\text{Vs/m}^2) = 2{\cdot}38 . 10^{-9} H^2 (\text{A/m})$$
$$= 1{\cdot}508 . 10^{-5}\,B^2 (\text{in gauss}) = 1{\cdot}508 . 10^{-5} H^2\,(\text{oerstd}).$$

At constant potential difference V and magnetic induction B the radius r increases with the square root of the ionic mass M. Conversely, at fixed radius r of the electromagnet, ions of increasing mass can be brought to focus at the exit slit consecutively by increasing the magnetic induction. For a radius of 25 cm and an electric potential difference of 2000 V we find that

$$M\,(\text{AMU}) = 1508\,B^2\,(\text{Vs/m}^2) \qquad (11.5)$$

Accordingly, the entire range of elements from $^1H^+$ to $^{238}U^+$ can be focused onto the exit slit by varying the magnetic induction from 0·025 Vs/m² (250 gauss) to 0·4 Vs/m² (4000 gauss).

An important feature of the mass spectrometer is the extremely deep vacuum required to permit unobstructed motion of the ions. Powerful vacuum pumps reduce the pressure to below 10^{-10} atm., and the various parts of the instrument are connected through high-vacuum seals.

The ions are collected by a conductor connected to earth through a large resistance ($10^{12}\ \Omega$). The small voltage over the resistance is amplified and measured with a potentiometer recorder. Detection is accurate, precise and very sensitive. Ionic currents of 10^{-15} A (10000 ions per second) can be measured, and this means that some elements can be determined down to 10^{-12} g with a relative precision of 1 per cent.

By varying the magnetic induction in a regular way a complete *mass spectrum* of the sample is obtained, where each deflection of the recorder pen (i.e. each peak) corresponds to ions of a particular mass.* The resolving power of a mass spectrometer is commonly defined as

$$R = \left(\frac{M}{\Delta M}\right)_{10\%} \qquad (11.6)$$

where ΔM is the mass difference such that the valley between two equally strong peaks of (nearly equal) mass M amounts to 10 per cent of the peak heights (compare this definition with Equation 3.4). Single-focusing mass spectrometers have typical resolving powers of about 2000, although instruments have been built with resolving powers as high as 10000.

* An alternative way to scan the mass spectrum is to vary the accelerating potential difference V (compare Equation 11.4).

An interesting application of thermal ionization mass spectrometry is *isotope dilution analysis*, which may be illustrated by the example of uranium. The natural composition of uranium is 99·274 per cent ^{238}U and 0·720 per cent ^{235}U. A minor variation of the magnetic induction around the value of 0·4 Vs/m^2 calculated from Equation 11.5 produces two uranium peaks in the mass spectrum with an intensity ratio equal to the natural isotopic abundancy ratio R_n. Now, if we add W g of *pure* uranium isotope ^{235}U to S g of the original sample, we change the isotopic abundancy ratio to the value R_m, which can also be measured with the mass spectrometer. If the original sample contained Q_{235} gram of ^{235}U and Q_{238} gram of ^{238}U, we have the following relations:

$$\frac{Q_{235}}{Q_{238}} = R_n = 0{\cdot}00725$$

$$\frac{Q_{235}+W}{Q_{238}} = R_m$$

$$\frac{Q_{235}+Q_{238}}{S} = G$$

and the weight fraction G of total uranium in the sample is found as

$$G = \frac{R_n+1}{R_m-R_n} \cdot \frac{W}{S} \tag{11.7}$$

In this equation W and S are known, while R_m and R_n can be measured. In this way uranium can be determined in geological samples down to concentrations of 10^{-8} per cent (10^{-4} ppm) with a relative precision of 1 per cent. Obviously, the method can be extended to other elements, provided that the natural composition of the element contains at least two isotopes and provided further that pure isotopes (or at least strongly enriched fractions) of the element are available. The precision of isotope dilution analysis ranges from 0·5 to 2 per cent.

11.2. Inorganic r.f. spark source mass spectrometry

The extent of thermal ionization used in the apparatus of Fig. 11.1 depends strongly upon the ionization energy of the sample element. Whereas alkali metals (with low ionization energy) are practically completely ionized, the ionization yield decreases rapidly to a value of 10^{-6} for ionization energies of 8 eV; elements with an ionization energy exceeding 9 eV (Be, Zn) are virtually not ionized. Thermal ionization is highly selective since it is much more efficient for one element than it is for another. Consequently, this source of ions is unsuitable for the simultaneous analysis of many elements in a sample.

A nearly non-selective source of ions is the *radiofrequency spark discharge* (Fig. 7.5a, p. 151), struck between two conductive electrodes placed in close proximity (separation less than one millimetre) in vacuum. Metallic samples can be shaped into electrodes and used as such, and non-conductive powdered samples can be placed in graphite electrodes. The discharge mechanism is not completely understood, but it causes all elements of the sample to become ionized to approximately the same extent. This makes the r.f. spark source the preferred source for the simultaneous mass spectrometric analysis of many elements in an inorganic sample.

There are, however, also disadvantages. In the first place, the spark discharge shows strong fluctuations, so that the mass spectrum cannot be scanned by varying the magnetic induction as was done in the preceding section. Instead, the magnetic induction is kept at a constant value and all ions are recorded simultaneously on a photographic plate (compare the direct current arc in § 7.2). Just as for photographic detection of electromagnetic radiation (§ 7.3), the calculation of ionic currents from photographic mass densities is cumbersome. In addition, the photographic emulsion is not equally sensitive for all ions. These factors limit the precision of quantitative analysis to about 25 per cent. Obviously, however, photographic detection offers the possibility of rapid qualitative and semi-quantitative estimation of many elements in a single exposure.

A second disadvantage is the influence of the sample and the discharge parameters (r.f. voltage, electrode gap) upon the ion yield. Systematic errors arising from differences between natural samples and artificial standards may amount to a factor of three in unfavourable cases.

The mass spectrum produced by the r.f. spark source is quite complex, because in addition to singly charged ions, double and triple ions are formed, as well as ions of molecular fragments. Appreciable resolving powers may be required to separate the individual peaks. For example, the ratio M/e that determines the position of a peak in the spectrum is nearly equal for $^{64}Ni^{2+}$ ($M/2e = 31{\cdot}9639$), $^{32}S^{+}$ (31·9720) and $^{16}O_2^{+}$ (31.9898), requiring resolving powers ($M/\Delta M$) of 4000 and 2000, respectively.

Now, resolving powers of this order can be obtained with good single-focusing mass spectrometers. However, there is another reason why this type of instrument cannot be used in conjunction with an r.f. spark ion source. In formulating Equation 11.1 it has been assumed that the ions have no initial velocity. This is permissible for thermal ionization, where the initial velocity is quite negligible. However, the ions produced in the r.f. spark have kinetic energies varying between 0 and 1000 eV. After acceleration by an electric potential difference of 10000 V, the final kinetic energies of ions of a single mass may differ by some 10 per cent. Since, according to Equation 11.3, the radius r is proportional to the square root of the kinetic energy $\frac{1}{2}Mv^2$, ions of similar mass will spread over various orbits in the magnetic field and give rise to very broad lines on the photographic plate.

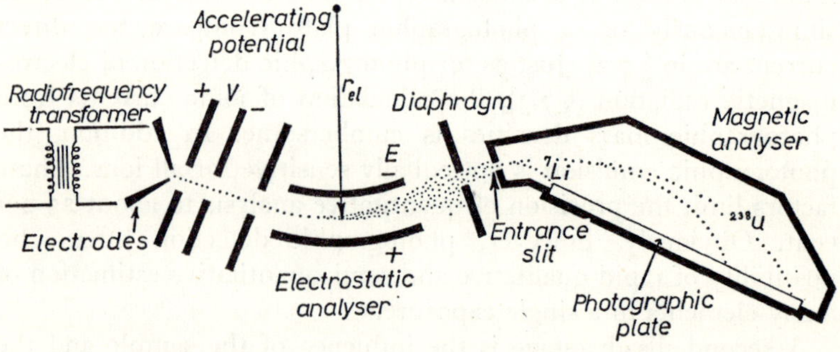

FIG. 11.2. Double-focusing mass spectrometer. Electric potential difference $V = 10000$ V. Electrostatic analyser over 32° with a radius $r_{el} = 20$ cm. Electric field strength $E = 10^5$ V/m. Magnetic analyser over 90°. Magnetic induction $B = 1$ Vs/m^2 (perpendicular to the paper). Radius r equal to 3·8 cm for $^7Li^+$ and 22 cm for $^{238}U^+$. Length of the photographic spectrum 25 cm.

Therefore, a *double-focusing* mass spectrometer must be applied, as shown in Fig. 11.2. The accelerated ions first enter an electric field with a field strength E, which deflects the ions according to the equation

$$Mv^2/r_{el} = eE$$

or

$$r_{el} = \frac{\frac{1}{2}Mv^2}{\frac{1}{2}eE} \qquad (11.8)$$

The electrostatic analyser separates the ions according to their kinetic energy $\frac{1}{2}Mv^2$. For fixed values of the radius r_{el} and the electric field strength E, the ions emerging from the exit slit of the electrostatic analyser have equal kinetic energy of $\frac{1}{2}eEr_{el}$. Substitution of this value in Equation 11.3 yields for the radius of the ionic orbit in the magnetic analyser,

$$r_{magn} = \sqrt{\left(\frac{Er_{el}}{B^2} \cdot \frac{M}{e}\right)} \qquad (11.9)$$

which is of a form similar to Equation 11.3.*

Commercially available double-focusing mass spectrometers reach resolving powers of up to 100000. Obviously, this requires very stable magnetic fields, but it has been shown in § 10.3 that these can be realized in modern instruments. Mass spectrometers of this type allow a very precise determination of ionic and isotopic masses. The important consequences for organic analysis will be discussed in § 11.7.

Very low limits of detection can be reached with r.f. spark source mass spectrometry. Some 10000 ions are sufficient to produce a just visible line on the photographic plate, and this corresponds to about 10^{12} atoms in the sample. Consequently, the absolute limit of detection can be as low as 0·1 ng. On the other hand, the total ionic charge must be limited to 1 μC or 10^{13} ions in order to restrict the continuum background intensity. This limits sample consumption to a few μg and the relative limits of detection are

* The operating principle of the double-focusing mass spectrometer is more complicated than is suggested by this simple derivation. In the magnetic analyser the ions are dispersed according to both their mass and their kinetic energy. The electrostatic analyser is adjusted to nullify the dispersion according to kinetic energy.

10^{-7} per cent or 10^{-3} ppm. For some elements, such as carbon, oxygen and hydrogen, the detection limits are only about 1 ppm, because it is very difficult to remove these elements completely from the mass spectrometer.

Inorganic mass spectrometry with the r.f. spark source is used for trace analysis and microanalysis. Special techniques have been developed for handling micro-amounts of samples such as semi-conductors, crystallites, biological tissues, etc.

11.3. Ionization of organic compounds

Neither thermal ionization nor r.f. spark ionization is suitable for organic substances. The ionization energy of C, H, O and N is much too high for thermal ionization and the precision of the r.f. spark source is much too low to replace classical methods of organic element analysis.

For the important application of mass spectrometry in organic analysis, another type of ion source is used, as shown in Fig. 11.3. A few milligrams of sample are introduced into a reservoir rarefied to a pressure of 10^{-8} atm. The sample is heated, and a small flow of sample vapour leaks through a filter into the high vacuum (10^{-10} atm.) compartment of the mass spectrometer. Alternatively,

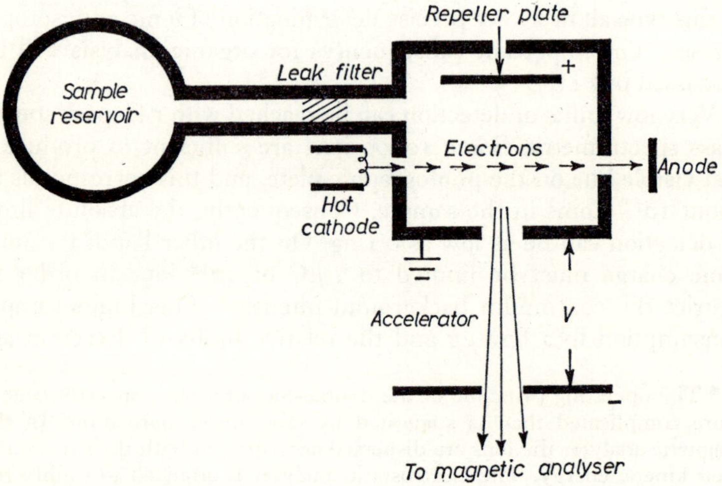

FIG. 11.3. Electron impact ion source for organic compounds.

the sample may be introduced directly into the high-vacuum system, but in all cases the vapour pressure of the sample must be large enough for the sample to volatilize sufficiently and produce a vapour.

The gaseous sample molecules are now bombarded with electrons emitted by a hot wire cathode. The energy of the electrons is usually about 70 eV, which is much more than the ionization energy of organic molecules, which is typically of the order of 10 eV. In consequence of this surplus energy the electron impact not only produces positive ions of the sample compounds, but may also rupture one or more chemical bonds of a molecule. The main processes occurring under the influence of electron impact are the following.

(*a*) *Single ionization* of a complete molecule according to

$$M + e \rightarrow M^{+} + 2e$$

The monovalent ion formed is called the *molecular ion* or the *parent ion*. Twofold ionization is a rare phenomenon in an electron impact ion source.

(*b*) *Cleavage* of one or more chemical bonds according to

$$R_1 - R_2 + e \rightarrow R_1 + R_2^{+} + 2e$$

where it should be remarked that multiple cleavage leading to two or more smaller fragments occurs much more frequently than single cleavage of only one bond. It should further be noted that any one of the fragments can carry the positive charge, although it is not unusual that the positive charge is transferred preferentially to one particular fragment. Negative ions, however, are seldom formed.

(*c*) In some cases the observed fragments can only be formed if the cleavage of a chemical bond is accompanied by the migration of a hydrogen atom from one part of the molecule to another part. An example of such *rearrangement* processes is the fragmentation of a ketone according to

$$\underset{\underset{\displaystyle O}{\|}}{RC}CH_2{-}CH_2R' + e \rightarrow \underset{\underset{\displaystyle OH}{|}}{RC}{=}CH_2^{+} + CHR' + 2e$$

(*d*) The above processes of ionization and fragmentation occur within a few nanoseconds inside the ion source chamber. Sometimes, however, an ion may split during its flight through the magnetic analyser. The parent ion of normal butane, for example, ejects a neutral fragment of mass 15 (CH_3), the positive charge remaining on the C_3H_7 part. In contrast to the other peaks in the mass spectrum, such *metastable ions* produce relatively broad peaks at non-integral mass positions.

As a result of this variety of processes a single organic compound does not yield one peak but a complete mass spectrum. As will be discussed in the following sections, each organic compound has a characteristic *fragmentation pattern* and a characteristic mass spectrum that can be used to identify the compound and, more important, to deduce its chemical structure.

The initial kinetic energy of the positive ions is small enough to permit the use of a single-focusing mass spectrometer. The ions are repelled by a positively charged plate towards the electrostatic accelerator and dispersed in the usual way (Fig. 11.1). Sample introduction, ionization and fragmentation are sufficiently stable to allow electrical detection and magnetic scanning of the mass spectrum. In modern instruments the electrometer detector shown in Fig. 11.1 is replaced by an electron multiplier, which operates like the photomultiplier discussed in § 4.3.

11.4. The mass spectrum of organic compounds

The mass spectrum of an organic compound ranges from a lowest mass number of about 15 to the molecular mass of perhaps several hundred atomic mass units. In a single-focusing mass spectrometer with a resolving power of 500, each peak represents a molecular fragment, *rounded to a whole number of atomic mass numbers* (except for metastable ions). These fragments may be recognized on the basis of *isotopic masses rounded to the nearest unit.* The overall intensity of the spectrum clearly depends upon the amount of sample being analysed, but for identification and structure determination only the *relative* intensities of the peaks need be known.

The mass spectrum is therefore reduced to the form shown in Fig. 11.4 for methanol. The strongest peak at mass number 31

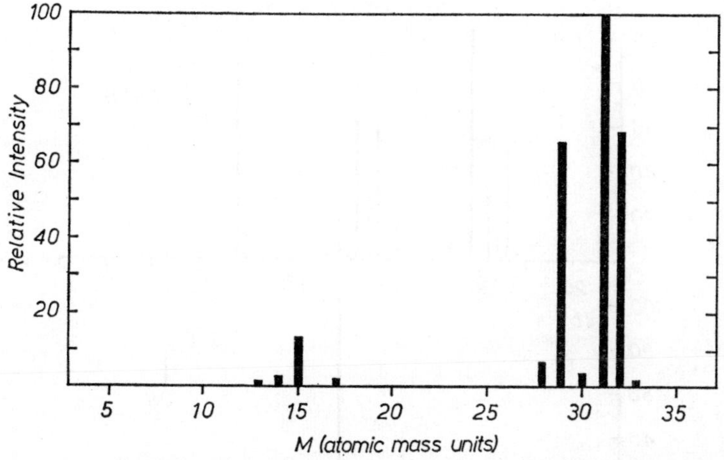

FIG. 11.4. The mass spectrum of methanol.

Fragment	*M/e*	*Relative intensity*	*Fragment*	*M/e*	*Relative intensity*
C	12	0·33	CHO	29	64
CH	13	0·72	CH_2O	30	3·8
CH_2	14	2·4	CH_3O	31	100 (base peak)
CH_3	15	13	CH_3OH	32	66 (molecular ion)
O	16	0·21	$^{13}CH_3OH$	33	0·98 $(M+1)$
OH	17	1·0	$CH_3{}^{18}OH$	34	0·14 $(M+2)$
CO	28	6·3			

(actually corresponding to a ratio M/e of 31 AMU and one unit of charge) is called the *base peak*, and is assigned a relative intensity of 100. The intensity of other peaks is expressed in proportion to the base peak. The complete mass spectrum is given in the table below Fig. 11.4. Very weak peaks cannot be represented in the figure, but are often extremely important for structure determinations.

The peak at mass number 32 is due to ionized methanol and is called the *molecular ion* peak. This is not always a strong peak. The examples in Fig. 11.5 show that in some mass spectra the peak of the molecular ion is completely missing, indicating that all molecules of the compound are disintegrated into smaller fragments in the electron impact source.

The peaks at masses smaller than 32 in Fig. 11.4 are now readily attributed to the various fragments resulting from ionization of

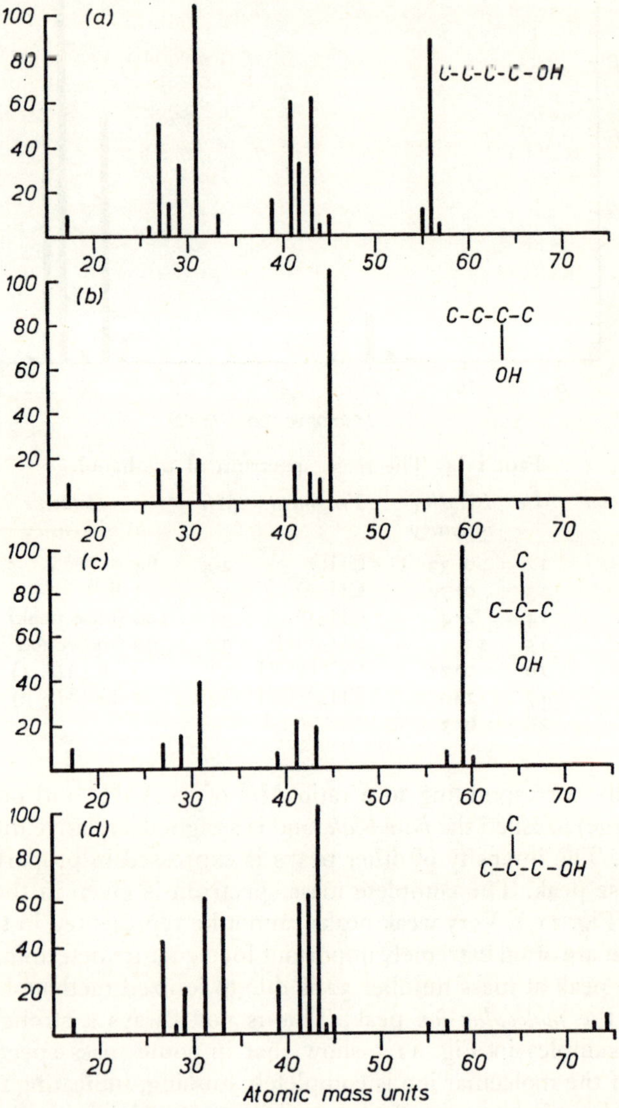

FIG. 11.5. Mass spectra of butanol isomers. (a) Butanol-1; (b) Butanol-2; (c) 2-methyl-propanol-2; (d) 2-methylpropanol-1. Note that only the last spectrum shows the peak of the molecular ion ($M = 74$).

methanol molecules. The peak intensities show that methanol predominantly ejects one or three hydrogen atoms. In addition there are moderately strong peaks corresponding to the CH_3-group and the CO-group. It should be noted that the CH_3-peak is much stronger than the OH-peak, showing that in the cleavage of the C–O bond the positive charge is preferentially transferred to the CH_3-group. The neutral OH-group is not deflected in the magnetic analyser, and only the positive ions are detected.

However, we also note in the spectrum of methanol a small peak at mass number 33, which is one mass unit heavier than the molecular ion. This is partly due to the fact that methanol ions may acquire one additional hydrogen atom. Such ion–molecule reactions sometimes occur. However, there is another reason for the appearance of peaks at one or two units of mass above the molecular ion, which is extremely valuable in structure determinations. It has been remarked in § 9.1 that many elements have different isotopes. For example, natural bromine consists of nearly equal proportions of ^{79}Br and ^{81}Br. Consequently, the mass spectrum of methyl bromide shows two nearly equally strong peaks at mass numbers 94 ($CH_3{}^{79}Br$) and 96 ($CH_3{}^{81}Br$), because one half of the molecular ions contains the lighter isotope and the other half the heavier isotope. In fact, the occurrence of two equally intense peaks at a separation of two mass units is highly characteristic of the presence of one bromine atom.

Of more general importance is the fact that carbon consists of 98·9 per cent ^{12}C and 1·1 per cent ^{13}C. Now, if the molecular ion peak of methanol with one carbon atom has a relative intensity of 66, we expect an additional peak at mass number 33 due to $^{13}CH_3OH$ with a relative intensity of $0{\cdot}011 \times 66 = 0{\cdot}72$. It is true that a methanol molecule with one hydrogen atom of mass 2 (deuterium) will also contribute to this peak at $M+1$, but natural hydrogen contains only 0·01 per cent of deuterium, and so this contribution is usually negligible.

The intensity of the peak at $M+1$ due to the presence of one ^{13}C atom will clearly increase with the number of carbon atoms in the molecule. If a compound contains two carbon atoms, the relative intensity of the peak at $M+1$ is $2 \times 1{\cdot}1 = 2{\cdot}2$ per cent of the molecular ion peak. In the spectrum of benzophenone the peak at mass number 183 is already 14 per cent of the molecular ion peak at 182. This indicates that benzophenone contains $14/1{\cdot}1 = 13$ carbon atoms ($C_6H_5COC_6H_5$). In this way peaks of smaller intensity provide important information, especially those at larger mass number.

11.5. Identification and structure determination

Just like the infra-red spectrum the mass spectrum of a substance is a characteristic 'fingerprint', by which the substance can be recognized. The components of a homologous series, such as normal alkanes, can be readily distinguished by their mass spectra, because the spectrum of heavier compounds will extend to greater mass numbers. Also, isomers give rise to substantially different mass spectra, as is shown by the spectra of four butanol isomers in Fig. 11.5. Yet the mass spectrum is not as frequently used for identification purposes as is the infra-red spectrum. The main reason for this is that the mass spectrum depends rather strongly upon the instrumental conditions (notably upon the properties of the ion source). Unknown compounds analysed with a particular instrument cannot be easily identified by comparison with reference spectra recorded on other instruments. Some laboratories have made it a practice to store each new mass spectrum and in this way build up their own library of reference spectra suitable for one particular instrument. One other difficulty is the extremely large number of peaks present in the mass spectra of heavier molecules. This makes it a tedious task to compare the spectrum of an unknown compound with even a few reference spectra.

An important application of mass spectrometry to qualitative organic analysis is in the determination of the chemical structure of an unknown compound from its mass spectrum. However, the mass spectrum offers no feature equivalent to the characteristic group frequencies in infra-red spectrometry or the chemical shift in nuclear magnetic resonance spectrometry (Chapters 5 and 10). It is, therefore, not so easy to formulate general rules and procedures to be followed in solving the problem posed by an unknown mass spectrum. Long experience, sound knowledge of organic reaction mechanisms and logical reasoning are essential tools in clarifying the mass spectra of larger molecules.

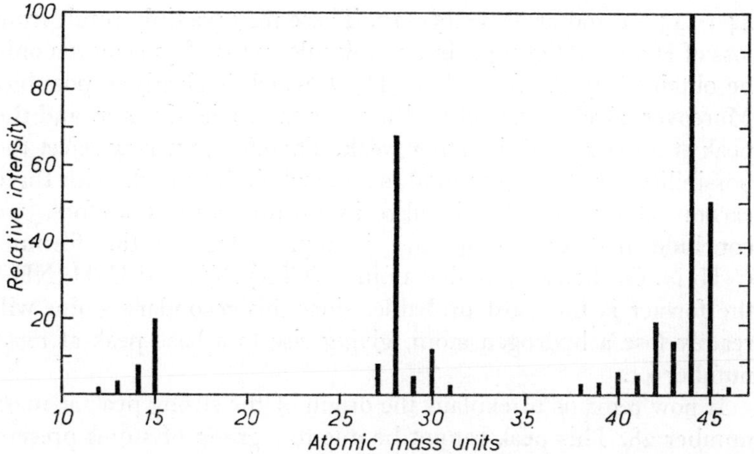

FIG. 11.6. Mass spectrum of an unknown compound (explanation in text).

First, the masses of the most stable fragments are noted in the mass spectrum, and the molecular mass is derived from the molecular ion peak (if present). Next, hypothetical compositions are formulated that agree with this molecular mass. As was discussed in the preceding section, the $M+1$ peak due to ^{13}C provides a clue to the number of carbon atoms in the molecule. This argument may be extended to the number of carbon atoms in stable fragments responsible for strong peaks at lower mass number. For small molecules (molecular weight less than 100) the number of possible compositions will be limited and some of these can possibly be rejected on the basis of other data. If a structural formula has been tentatively formulated, the reverse procedure is to explain the observed mass spectrum on the basis of the proposed molecular structure. This final step is often the most difficult one and it requires a deep insight into possible cleavage reactions and rearrangement processes of the proposed molecule.

A very simple example is the mass spectrum in Fig. 11.6. Of the many strong peaks let us assume that the one at highest mass number (45) is due to the molecular ion. Because the peak at mass number 46 is about 2·2 per cent of the peak at 45, the molecule must contain two carbon atoms. Strong peaks also occur at

$44 = 45-1$ and at $28 = 45-17$. These may possibly result from loss of H and OH groups, but a molecular mass of 45 can then only be obtained by the formula C_2H_5O, which is clearly impossible. Moreover, alcohols usually release one molecule of water, and the peak at $27 = 45-18$ is rather weak. Therefore, we may reject the possibility that the compound is an alcohol. A formula with three carbon atoms (C_3H_9) is also impossible and, therefore, we conclude that the compound is represented by the formula C_2H_7N. Of the two possible amines, $(CH_3)_2NH$ and $C_2H_5NH_2$, the former is the most probable, since this secondary amine will readily lose a hydrogen atom, giving rise to a base peak at mass number 44.

It now remains to explain the origin of the strong peak at mass number 28. This peak cannot be due to a group of atoms present as such in the molecule. It is probably due to C_2H_4, which may be obtained from dimethylamine by subtracting one molecule of ammonia (NH_3). This is only possible if two bonds in the original molecule are broken and new bonds formed. This is an example of the *rearrangement* processes mentioned in § 11.3. In the present example the rearrangement process is rather a nuisance, but in other cases the opposite may be true. Double bonds, for example, frequently lead to characteristic, well-known rearrangements that can be recognized by the experienced mass spectrometrist and will assist him in interpreting the mass spectrum.

11.6. Coupling of a mass spectrometer to a gas chromatograph

The above examples have made it clear that in order to determine its structure, the unknown compound must be supplied to the mass spectrometer in pure form. In contrast to ultra-violet, infrared or NMR spectrometry there are no harmless impurities in mass spectrometry. Each impurity is an interference that contributes unwanted peaks to the mass spectrum. Their absence is the more important, in that the smaller peaks in the mass spectrum often provide important information.

Fortunately (and this is one of the main reasons for using this method of analysis), mass spectrometry is so sensitive that several chromatographic techniques can be used to separate the components of a mixture. Whereas infra-red spectrometry requires at least several micrograms (§ 5.8) and nuclear magnetic resonance spectrometry even more than that (§ 10.6), it is quite feasible to obtain a good mass spectrum from less than $0{\cdot}1\ \mu g$ of gaseous sample. In fact, the lower limit to the sample size is usually not the sensitivity of the instrument, but our ability to manipulate minute amounts of sample. This makes mass spectrometry the method of choice for the spectrometric identification of components separated by chromatographic methods.

Now, obviously, a particular component may be collected after the chromatographic separation is completed, and transferred subsequently to the mass spectrometer. If thin-layer chromatography is used, this procedure is necessary. However, in the case of gas chromatography the separated components are in the gaseous state and are suitable for direct introduction into a mass spectrometer. If the mass spectrometer were coupled directly to the outlet of the gas chromatographic column, it would be possible to identify each peak on the recorder of the gas chromatograph from its corresponding mass spectrum.

This exciting possibility has attracted much attention over the past ten years, and the inherent difficulties have been gradually solved. The main problem is that the sample component eluted from the gas chromatograph is diluted by some thousandfold excess of helium carrier gas at atmospheric pressure, whereas the ion source of the mass spectrometer operates at a pressure of less than 10^{-8} atm.

This problem may be overcome by simply splitting the gas flow and admitting only about 10 per cent into the ion source. Although the main stream can now be used to record the gas chromatogram in a conventional way, this solution has the disadvantage that from the point of view of mass spectrometry most of the sample is wasted. Therefore, it is commonly preferred to utilize a molecular separator between the gas chromatograph and the mass spectrometer, where the lighter helium atoms are separated from the heavier organic molecules, usually through a diffusion mechanism.

Another requirement is that the scanning rate of the mass spectrometer must be increased appreciably, so that the entire mass spectrum is recorded within the few seconds that the component is eluted from the gas chromatograph. Note that at this scanning rate each mass unit is passed in a few milliseconds! Naturally, this makes high demands upon the detection system. The conventional pen recorder must be replaced by a galvanometer recorder. An interesting alternative development is to connect the detector of the mass spectrometer to a high-speed computer which converts the detector output into a table of mass units with the appropriate peak intensities.

11.7. High-resolution mass spectrometry

The commonly-used chemical atomic masses of the elements are based upon a scale of 12 AMU assigned to the isotope ^{12}C. Since many elements are actually a mixture of different isotopes (compare chlorine), the atomic masses are generally not equal to whole numbers. However, the mass of an individual isotope also differs slightly from an integral number of AMU. The exact masses of the most abundant isotopes of carbon, hydrogen, oxygen and nitrogen

TABLE 11.1. Determination of the elementary composition from the exact mass.

Substance	*M* in atomic mass units	ΔM	$\frac{M}{\Delta M}$
^{1}H	1·00783		
^{12}C	12·00000		
^{14}N	14·00307		
^{16}O	15·99491		
$C_4H_4O_2$	84·02113		
		0·01123	7480
$C_3H_4N_2O$	84·03236		
		0·01123	7480
$C_2H_4N_4$	84·04359		
		0·01392	6040
C_5H_8O	84·05751		
		0·01124	7470
$C_4H_8N_2$	84·06875		
		0·02515	3340
C_6H_{12}	84·09391		

are presented in Table 11.1. These values have been determined with a double-focusing mass spectrometer, and are seen to differ slightly from the nominal, rounded values (except for ^{12}C, of course).

Consequently, the mass of a molecule made up of these four elements also differs slightly from a whole number of AMU. Table 11.1 presents six possible (if somewhat hypothetical) molecules consisting of carbon, hydrogen, oxygen and nitrogen, all with a nominal mass of 84. In a single-focusing mass spectrometer of moderate resolving power each substance gives rise to a molecular ion peak at mass number 84. However, a resolving power of 10000, as offered by a double-focusing mass spectrometer, is clearly sufficient to distinguish between these substances. For example, if the exact mass is found to be 84·058, this would be unambiguous proof that the peak is due to a substance (or fragment) with formula C_5H_8O.

If the mass numbers to be assigned to the peaks in a mass spectrum need be known to the nearest whole number only, the scale can be calculated from Equation 11.5. If, however, chemical formulae are to be deduced from exactly determined non-integral mass values, the scale must be established very precisely. This can only be done by recording the well-known mass spectra of reference substances. Fluorolube (§ 5.4) is widely applied, since the mass spectrum of this substance shows many peaks.

Since modern double-focusing mass spectrometers can reach resolving powers up to 100000, precisely measured masses up to nominal values of $M = 500$ can be used directly to give the elementary compositions. Naturally, the number of possible formulae increases rapidly with increasing mass number. Again, computers may be used to compare the experimentally measured exact mass with all possible combinations of C, H, O, N, P, Cl, etc. atoms, the exact masses of which can be stored in the computer memory.

11.8. Quantitative organic mass spectrometry

As remarked in § 11.5, the mass spectrum of a pure compound depends rather strongly upon the instrumental conditions. However, for a particular mass spectrometer, and at constant operating conditions of the electron impact ion source, the relative distribution of the peak intensities is quite reproducible. The overall intensity of the spectrum is proportional to the partial pressure of the substance in the sample reservoir (Fig. 11.3). In a mixture of organic compounds the partial pressures are directly proportional to the molar concentrations, in consequence of the low total pressure utilized in a mass spectrometer. As a result mass spectrometry can be used for the quantitative analysis of organic mixtures, where the concentration of a component can be derived from the intensity of a peak in its mass spectrum.

However, there are two complications. In the first place, the mass spectrometer is not equally sensitive for all compounds, so that the observed peak intensities must be multiplied by a sensitivity factor derived from the spectrum of a pure compound. Such sensitivity factors also depend upon the operating conditions and must be calibrated from time to time.

The second problem is an obvious one. Since each component of a mixture gives rise to a complete mass spectrum, it will be difficult to find a peak that is specific for a single component. In general, several components of a mixture contribute to the total intensity of a selected peak, which can be written as

$$m_i = x_P\, a_{P,i}\, s_P + x_Q\, a_{Q,i}\, s_Q + \ldots \qquad (11.10)$$

Here m_i is the total intensity measured at mass number i, x_P is the molar fraction of component P, $a_{P,i}$ is the relative intensity of the peak at mass number i in the mass spectrum of the pure component P, and s_P is the sensitivity factor for component P. Consequently, in analysing a mixture the procedure of multicomponent analysis must be applied, as explained in § 4.6, where the number of measured peaks is at least equal to the number of components in the mixture. This form of quantitative analysis will, therefore, not be contemplated for simple mixtures, where competitive methods (notably gas chromatography) are more convenient.

For some special cases of quantitative organic analysis the mass spectrometer is indispensable. The high sensitivity of the mass spectrometer is essential in the analysis of gaseous traces in high-vacuum systems or in leak testing. The highly complex mixtures of natural hydrocarbons (petroleum) can be analysed by mass spectrometry, because some mass peaks are characteristic of distinct classes of hydrocarbons (alkanes, aromatics, cycloalkanes, monoalkenes, etc.).

Bibliography

1. MORRISON, G. H. (ed.), *Trace Analysis, Physical Methods*, Interscience, New York, 1965.
2. AHEARN, A. J. (ed.), *Mass Spectrometric Analysis of Solids*, Elsevier, Amsterdam, 1966.
3. McLAFFERTY, F. W., *Interpretation of Mass Spectra* (*An Introduction*), Benjamin, New York, 1967.
4. HILL, H. C., *Introduction to Mass Spectrometry*, Heyden and Son, London, 1966.
5. BEYNON, J. H., SAUNDERS, R. A., and WILLIAMS, A. E., *The Mass Spectra of Organic Molecules*, Elsevier, Amsterdam, 1968.
6. WILLIAMS, D. H., and FLEMING, I., *Spectroscopic Methods in Organic Chemistry*, McGraw Hill, London, 1966.
7. LIENITZ, H., *Massenspektrometric*, Verlag Chemie, Weinheim, 1968.
8. ROBOZ, J., *Introduction to Mass Spectrometry*, Interscience, New York, 1968.

Problems

1. Adapt Equations 11.1 and 11.4 to the case of doubly or triply ionized atoms. Explain the near coincidence of $^{32}S^{2+}$ and $^{16}O^{+}$ in the mass spectrum. Calculate the resolving power required for clear separation of the peaks due to these ions.

2. How would Equation 11.7 change if the pure uranium isotope added to the sample is not ^{235}U but ^{238}U? Which alternative would you prefer?

3. The natural composition of indium is 4·23 per cent ^{113}In and 95·77 per cent ^{115}In; 9·51 μg of pure ^{113}In is added to a 1 g sample. The intensity ratio of ^{113}In and ^{115}In is now measured to be 1 : 9·75. What is the indium content of the sample?

4. The direct determination of indium, as in problem 3, is impossible in the presence of cadmium, since the peak of ^{113}Cd coincides with that of ^{113}In. Obviously, either the indium or the cadmium must be separated out, possibly by extraction from a solution of the sample.

(*a*) If indium is extracted *after* addition of ^{113}In, must the extraction be quantitative?

(*b*) What is your answer if indium is extracted *before* addition of ^{113}In?

(*c*) What is your answer if cadmium is extracted?

(*d*) Could you suggest a method of determining indium in the presence of cadmium, not requiring separation? (Hint: look up the isotopic composition of natural cadmium.)

5. A gaseous substance consisting of C, H, O or N shows a peak at 28·03 mass units. What is the substance?

6. Considering the remark in § 11.4 about bromine, how would you detect the presence of a chlorine atom from the mass spectrum of a compound?

7. The highest mass number peaks of a gaseous compound are found at 64 (base peak), 65 (relative intensity 2·2), 66 (rel. int. 33) and 67 (rel. int. 0·7). Considering the previous question, can you deduce the structure of the compound?

8. The peak at mass number $M+2$ in the spectrum of methanol is due to molecular ions containing the oxygen isotope ^{18}O. Calculate the isotopic abundance of oxygen from the intensities given in the caption to Fig. 11.4.

9. The mass spectrum of a substance shows a base peak at mass number 78. The only peak at higher mass is at 79, with a relative intensity of 6·5. What is the substance?

10. The mass spectrum of benzophenone shows the following strong peaks:

mass number	77	105	182
relative intensity	62	100	60

What fragments are responsible for these peaks? What intensity would you predict for a peak at mass number 106?

11. The exact molecular mass of a compound consisting of carbon, hydrogen, oxygen or nitrogen is $58{\cdot}042 \pm 0{\cdot}001$. What is the substance?

12. Show that a molecular ion peak of odd mass number indicates the presence of one (or three, five, etc.) nitrogen atoms in the molecule.

CHAPTER TWELVE

CONCLUDING REMARKS

12.1. Selecting a method of analysis

This book has been concerned with *optical methods of analysis* utilizing electromagnetic radiation for the qualitative characterization and the quantitative determination of one or more constituents of a sample. These are also *instrumental* methods of analysis, in the sense that an instrument other than a balance or a burette is used to obtain the analytical result. The following arguments may be advanced to explain the still growing popularity of instrumental methods of analysis:

(*a*) The rapid development of electronic equipment has greatly improved the reliability and simplicity of analytical instruments, whereas the price has been reduced, at least in comparison with the sharply increased costs of labour.

(*b*) Instrumental methods of analysis allow of more constant precision and more objective results.

(*c*) Instrumental methods have tremendously reduced the time and effort of obtaining results (chemical structure, trace analysis), and have provided answers that were otherwise impossible to get (crystal structure).

(*d*) An instrumental method of quantitative (and eventually also qualitative) analysis lends itself to mechanization and automation, with an inherent increase in speed and analysis rate, and at reduced cost per analysis, the expensive equipment notwithstanding.

With the increasing variety of instrumental methods of analysis available the analyst is confronted with the task of selecting the proper method for the problem in hand. Evidently, his choice will be restricted to the instruments available in his laboratory. However, facilities are rapidly expanding and it is certainly no exception to find all methods discussed in this book available in an analytical laboratory. Also, the possible acquisition of a new, sometimes expensive instrument must be carefully considered and justified.

Naturally, the analyst's choice is not limited to the (optical) methods discussed in this book. An isolated determination can often be performed with less preparation or higher accuracy by 'classical' means. In addition there are numerous problems where optical methods fail completely. For example, chromatographic methods will be indispensable in the analysis of complicated mixtures of organic compounds, even though spectrometric methods may be called upon for the identification or quantitative assessment of the separated components. But even if the nature of the problem points to an optical method of analysis (chemical structure, elemental composition), it will not always be a simple matter to decide which one of the methods discussed in this book is to be preferred.

In general then, problems of quantitative analysis may be attacked by a variety of methods, optical as well as non-optical. *The analyst should then select that method which will give him the desired answer for the problem in hand.* This statement implies that the analyst must be guided by the nature of the sample and by the demands imposed upon the determination. He need not seek a general method of great flexibility unless this is required by a wide range of sample compositions. By the same token he need not aim for a higher precision or a greater speed than necessary. As a simple example, let us consider the determination of manganese in a solid sample, and let the choice be limited to the optical methods considered in this book.

(*a*) We are asked for a rapid survey of the composition, and a semi-quantitative estimate of the manganese content of a single sample. The electric arc discharge gives us all metallic constituents with concentrations exceeding 10^{-4} per cent. For still lower concentrations, spark-source mass spectrometry may be used. The answer is known within one hour.

(*b*) A fairly accurate determination of manganese must be made for a few readily soluble samples. Atomic absorption flame spectrometry is capable of rapid analysis for concentrations over 0·001 per cent. Lower concentrations may be determined by means of molecular absorption spectrometry, but more effort may be required to remove interferences. Activation analysis is extremely sensitive, but requires radiation facilities and much more preparation.

(*c*) The number of samples is still limited but the precision must be very high. Photometric titration (through permanganate) might provide a solution for low manganese contents (and higher concentrations can be titrated visually).

(*d*) We are asked for a rapid analysis of a large number of similar samples, as for example in the steel industry. It is then worthwhile to spend more money and time on preparation, and look for methods that can handle the samples on a routine basis and possibly automatically. Spark spectrometry and X-ray fluorescence spectrometry are excellent alternatives. The former is preferable if, in addition to manganese, a large number of other elements must be determined. The latter is preferable if a high precision is required.

This example shows that in selecting a method of analysis several factors must be taken into account. Important criteria are:

(*a*) The nature and composition of the sample, on which depends the extent of sample treatment prior to the determination proper (dissolving the sample, removing interferences, preconcentration, etc.).

(*b*) The speed of analysis required.

(*c*) The precision and accuracy required.

(*d*) The amount of sample available, and the concentration level of the constituent to be determined.

(*e*) The number of samples, and the variability of the sample composition.

(*f*) The cost per analysis, especially for larger numbers; related to this is the possibility of automation.

Table 12.1 (p. 268) is an attempt to present a general survey of these aspects for the methods of analysis discussed in this book. It goes without saying that such a survey must be restricted to general features and broad outlines. Exceptional situations may be encountered in practice where generally suitable methods fail and less obvious methods must be resorted to. Moreover, many ingenious techniques have been described in the literature to extend the normal range of applicability of a method. The table does not take account of such 'exceptions to the rules'.

The reader is advised to read the explanation of the table in the following sections, before studying the table (and the notes to it) in more detail.

12.2. Table 12.1; general aspects

Optical methods of analysis may be divided in two broad categories. The first group includes methods for determining *individual elements* in the sample without any structural information. *Destructive* methods of analysis, using electric discharges or flames that consume and destroy the sample, obviously belong to this group. In principle, the sample may be recovered unaffected after neutron-activation analysis or X-ray fluorescence spectrometry. This is important for rare or expensive samples, such as museum pieces. Usually, however, *non-destructive* analysis is a rare phenomenon, because the sample is frequently subjected to chemical treatment to reduce systematic and random errors.

The second category embraces those methods that in the first instance determine *molecules* and, in principle, provide structural information about sample constituents. In so far as such molecules can be indicative of the presence of a particular element, molecular absorption spectrometry may be classified under both headings. Also, mass spectrometry can be used either for the determination of individual elements (with the r.f. spark source) or for the determination of molecular structures.

One important factor to be considered before introducing a new method is the price of the instrument. Now, for each method, instrument prices vary considerably, since they reflect the specifications of the instrument. Higher speed, improved accuracy or greater resolving power must clearly be paid for. The table, therefore, distinguishes three categories. All prices include the *necessary* accessories, such as gas flow regulators, optical cells, microphotometer for handling photographic plates, etc. *Optional* accessories, such as a fluorescence accessory to ultra-violet spectrometers, are not included. A noteworthy practical rule states that about 20 per cent of the instrument price should be reserved each year for spare parts, service and maintenance, chemicals, recorder chart paper, etc. Depreciation may take another 10 to 20 per cent per year.

The cheapest category includes instruments of limited versatility. Here, the flame spectrometer or the fluorimeter will be equipped with filters instead of a monochromator, and the X-ray diffraction spectrometer operates in air and uses photographic films.

The normal category includes instruments for general use. The γ-spectrometer has a scintillation detector, the nuclear magnetic resonance spectrometer measures proton resonance at 60 MHz, and the X-ray fluorescence spectrometer scans the spectrum.

The most expensive category includes instruments for advanced research and for automatic routine analyses. The γ-spectrometer now has a semiconductor detector, the fluorimeter is equipped with two monochromators, the grating infra-red spectrometer extends to very large wavelengths, and the ultra-violet/visible spectrometer is a double-beam instrument. Spectrometers for spark discharge, X-ray fluorescence and nuclear magnetic resonance include a small computer* to convert the electric signals to (corrected) concentrations or for ensemble averaging of spectra. Advanced mass spectrometers (double-focusing, high resolution) and nuclear magnetic resonance spectrometers (220 MHz) may cost up to £70 000.

Another general feature is the degree of sophistication of a method. The table distinguishes between, on the one hand, the practical experience required to handle the instrument and to perform ordinary analyses, and, on the other hand, the experience necessary for introducing the method and adjusting it to novel conditions. In the latter respect, instruments for rapid and automatic routine analysis present the biggest problems. The inter-

* Alternatively, the price includes an *interface* to connect the instrument to a large computer.

pretation of infra-red, proton resonance and mass spectra requires not only experience but also a sound knowledge of the theoretical background. Complicated electronic equipment also requires the presence nearby of a specialized service-department (unless we want to depend upon the service of the manufacturer; but this clearly takes time).

Finally, the upper part of Table 12.1 contains some data about the sample, again related to the normal situation. For example, it is certainly possible to analyse solutions with an arc discharge, but in the usual arrangement the sample is introduced in solid form. The normal amount of sample is specified to indicate the amount that will provide an answer without trouble. However, in those cases where very little sample is available, it is important to know the minimum amount required for a (single) determination. Frequently, the lower limit is determined by the instrumental technique, but, as stated above, specially developed microtechniques are not considered in the table. It may also happen that the lower limit depends upon our skill in handling a tiny amount of sample. The mass spectrometer and, to a lesser extent, the spark spectrometer need only very little sample, but it is extremely difficult to manipulate a few micrograms of solid sample, and it is virtually impossible to shape it into an electrode. By contrast, even less than a microgram of gaseous sample (for example from a gas chromatograph) can be introduced into the ion source of a mass spectrometer. Finally, it must be recognized that an analysis of a very small sample may reveal only the major constituents. This is true, for example, for a d.c. arc exposure of a 1 μg sample.

12.3. Table 12.1; qualitative aspects

Arc spectrometry is most widely applied for the detection of metallic constituents in inorganic or organic samples. R.F. spark source mass spectrometry is more sensitive, but also more expensive. X-ray fluorescence is rarely used for qualitative analysis. Gaseous elements (rare gases, halogens) and anions can usually not be found by spectrometric methods. X-ray diffraction can be used to recognize crystalline substances. All these methods work well with mixtures.

	Ultra-v Absorption	
GENERAI		
Price	$0{\cdot}2^b$	—
(necess	2^b	2^c
access	—	10^c
Exper	Weeks	
	Months	
Specia	No	
Sampl	Solution	
	Org./inorg.	
	10 ml	
	0·1 ml	
QUALITA		
Detern	—	
Chemi	Org. subst.	
Identi	—	
Purity	Incidentally	
QUANTIT		
Compo	All ions, many molecules	
Optim (sta	0·3	
Interfe	Often	
	—	
	Masking, separation	
Limit	0·001–0·1[q]	
	10–1000	
Time	Days	
	Minutes[t] to hours[u]	
	1 minute	

he element sough
levels of 0·1 per
ower limits of det

elemental analysi
es the limit of det
ent. Precision in c

erences.

TABLE 12.1 Characterization of optical m

Method		γ-spectrometry	An
GENERAL ASPECTS			
Price in £1000 (necessary accessories included)	limited	—	
	normal	7	
	advanced	20	15
Experience required for	1. Operation	Month	Mo
	2. Setting up method	Year	Ye
Specialist required for maintenance		Yes	Some
Sample	physical state	Solid/liquid	Sol
	chemical state	Org./inorg.	Inorg
	usual amount	1 g	1 g
	minimum amount	Arbitrary	Arbi
QUALITATIVE ASPECTS			
Determination of elements			Not
Chemical structure (pure compound)			—
Identification (pure compound)			—
Purity control			Inorg
QUANTITATIVE ASPECTS			
Components to be analysed		Most elements	Meta elem
Optimum precision (standard deviation in per cent)		2	5
Interference	gross errors	Often	Nev
	systematic errors		Often
	to be counteracted by	Separations	Buffe
Limit of detection	relative in ppm	10^{-6}–0.1	1–1
	absolute in ng	0.01–100	1–1
Time needed for	setting up procedure	Weeks	We
	Sample pretreatment	Minutes to days	5 min
	Measuring	10–60 minutes	2 hou

a. Data refer to powder diffraction: the determinati tures requires different techniques and more ex (£25 000, small dedicated computer included).
b. Single-beam instruments.
c. Double-beam instruments.
d. Small computer or interface to large computer-in
e. Sample must be volatilized prior to analysis.
f. Alloys can be used directly as electrodes, an evaporated.
g. Data refer to organic samples; spectrum of inorgan larger amounts, although only a small fraction is

Infra-red spectrometry is the accepted method for the *identification* of organic compounds. Low-level impurities present no big problems, but complex mixtures are best avoided. The determination of the chemical structure of an unknown organic compound generally requires the compound to be isolated in pure form. Infra-red, NMR and mass spectrometry (possibly also ultra-violet) will be used *in combination*.

Whereas optical methods are indispensable for element detection and structure determinations, other methods may be considered for identification and purity control. Classical data, such as melting point or refractive index, should be judged with extreme care, however. Chromatographic methods (gas chromatography, thin-layer chromatography) have limited use for identification, but are often preferred for purity control, although very low-level impurities may escape detection. Extremely low concentrations of other constituents can sometimes be detected with a mass spectrometer. Related compounds (homologues) may not be distinguished by their infra-red spectra, and either NMR or mass spectrometry can then be helpful.

12.4. Table 12.1; quantitative aspects

The bottom part of Table 12.1 deals mainly with the quantitative determination of one or more constituents in a composite sample. Nuclear magnetic resonance spectrometry is rarely used for this purpose, and infra-red spectrometry or X-ray diffractometry is used only in special cases. For mass spectrometry the data refer to inorganic analysis with the r.f. spark ion source.

It must be emphasized from the outset that most instrumental methods of analysis and certainly all optical methods discussed in this book are *relative methods* when applied to quantitative analysis. In all cases concentrations are read from an analytical curve prepared and periodically checked with standards of known composition. This implies that the results (precision, accuracy) can never be better than the standards used. Indeed, it has been mentioned in several places throughout this book that a major problem in applying a particular method of analysis may be the standards of reference. This is perhaps obvious for methods prone to systematic errors, such as X-ray fluorescence and spark spectrometry. However, even in the apparently simpler cases of molecular absorption and flame spectrometry, the labels attached to ordinary laboratory chemicals should not be taken at face value. Notably, the possibility of an unknown water content in what appears to be a perfectly dry substance should not be overlooked.

This also seems a good place to express a warning. Many chemicals can be obtained in 'spectroscopically pure' quality. It should be realized that this label must be taken quite literally, and only confirms the absence of constituents that can be detected by spectrometric means. For example, a metal oxide with a guaranteed emission spectrometric purity of 99·999 per cent indeed contains no more than 10 ppm of other metallic elements, but the presence of excess oxygen is quite possible. In fact, the metal oxide content may be no more than 90 per cent of its nominal value. In the same way, spectrally pure solvents for absorption spectrometry are not necessarily 100 per cent pure, but are guaranteed to have excellent transparency over a certain wavelength range (ultraviolet or infra-red). For example, spectrally pure cyclohexane will be scrupulously free of benzene.

Now, let us consider the data in the bottom section of Table 12.1. The *precision* of a determination is quoted for optimum conditions, good instruments, and concentration levels well above the limit of detection. Consequently, the figures represent the optimum values attainable without application of special techniques.

Whereas the precision of an analytical result can, in principle, be improved indefinitely by mere repetition of the analysis, the *accuracy* is a much more critical quantity which is often limited by *interferences* due to the presence of other constituents in the sample. The table distinguishes between gross interferences that make the determination completely impossible, and minor interferences that lead to (still unacceptable) systematic errors. The borderline between the two categories is rather arbitrary.

Really serious interferences occur when the signal is overlapped by signals from other constituents. This is a rare phenomenon in flame spectrometry, but occurs fairly often in activation analysis and ultra-violet/visible absorption spectrometry. Sample pretreatment, such as masking or separation procedures (ion exchange, extraction) are then imperative, though time-consuming. This may be unacceptable for routine analysis, but it is seldom a problem for a few isolated determinations. Procedures for the removal of interferences can usually be found in the literature.

Gross interferences of this type never occur in arc/spark spectrometry, mass spectrometry or X-ray fluorescence spectrometry. By contrast, however, direct determinations by these methods are usually subject to systematic errors ranging from a factor of two (arc discharge, r.f. spark, mass spectrometry) to less than 25 per cent (spark discharge, X-ray fluorescence). Flame spectrometric analysis should always be checked carefully for the absence of systematic errors.

There are several procedures for overcoming systematic errors. Standard solutions in flame spectrometry are usually prepared so as to agree with the overall composition of the sample solution. In arc spectrometry or X-ray fluorescence spectrometry the sample is strongly diluted (buffering) to reduce the influence of the sample upon the measured intensities. Obviously, this raises the lower limit of detection. In spark spectrometry and X-ray fluorescence spectrometry the instruments are calibrated by means of reference samples of known composition, which can either be purchased or analysed by chemical methods. This is only profitable for routine analyses.

The (lower) limit of detection depends on many factors, notably the quality of the instrument, the nature of the constituent and, perhaps most important, the experience of the analyst. The ranges given in the table are conservative estimates which should be attainable for most constituents with a good instrument, following general procedures. A distinction should be made between *relative limits of detection*, referring to *concentration* levels (1 ppm = 10^{-4} per cent), and *absolute limits of detection*, referring to the absolute amount of a constituent that yields a detectable signal. As most samples are in the solid state, dissolving the sample produces a large dilution, and a corresponding change in detection limit. For example, a limit of detection of 0·1 ppm in solution normally corresponds to a limit of detection of 10 ppm with respect to the solid sample.

Generally, it takes several months to adjust and calibrate a spark spectrometer or an X-ray spectrometer for a new type of analysis. By contrast, procedures for flame spectrometry or visible absorption spectrometry can often be taken directly from the literature. Somewhere between these two extremes we find arc spectrometry and γ-spectrometry. Sometimes, standard procedures provide an answer within a few days, but more preparation is usually required to obtain optimum results with complicated samples.

Sample pretreatment is widely variable. Solid, homogeneous samples may be sparked directly, and liquid samples may be introduced as such into the flame. In other cases, however, solid samples may take a long time to dissolve (rocks). Separation procedures, masking or buffering, often required to remove interferences, will take at least a few minutes, but sometimes several hours are needed (ion exchange). The irradiation of the sample in activation analysis can be considered as sample pretreatment, and in unfavourable cases this must be prolonged for hours or even days.

By comparison, the actual measuring time is usually very short. On the other hand, the conversion of raw data to concentrations can be time-consuming. In the simplest case it involves only a reading from a graph (analytical curve in flame spectrometry or ultra-violet/visible absorption spectrometry). In spark spectrometry and X-ray fluorescence spectrometry routine analyses are usually handled by a high-speed computer. Arc spectrometry and inorganic mass spectrometry use photographic detection, and post-analysis processing may take several hours. In activation analysis and X-ray spectrometry the measuring time (counting time) depends upon the strength of the peak and the desired precision. Normally, a few minutes are sufficient, but occasionally it may amount to one hour.

APPENDIX I

PHYSICAL CONSTANTS

The internationally recommended system of units is the SI-system based on the metre, the kilogram, the second and the ampere, as basic units. The Gaussian system, based on the centimetre, the gram, the second and the electrostatic unit of charge (e.s.u.), is often used in the literature. The following table presents the values of some physical constants in both systems.

		Rational system		*Gaussian system*	
Symbol	*Description*	*Value*	*Unit*	*Value*	*Unit*
c	Vacuum velocity of light	$2{\cdot}998 . 10^{8}$	m/s	$2{\cdot}998 . 10^{10}$	cm/s
e	Electronic charge	$1{\cdot}602 . 10^{-19}$	C	$4{\cdot}803 . 10^{-10}$	e..s.u
m	Electronic mass	$9{\cdot}109 . 10^{-31}$	kg	$9{\cdot}109 . 10^{-28}$	g
AMU	Atomic mass unit (of $\frac{1}{12}$ ^{12}C)	$1{\cdot}660 . 10^{-27}$	kg	$1{\cdot}660 . 10^{-24}$	g
N	Avogadro's number	$6{\cdot}025 . 10^{-23}$	mol^{-1}	$6{\cdot}025 . 10^{-23}$	mol^{-1}
k	Boltzmann's constant	$1{\cdot}380 . 10^{-23}$	$J/^{0}K$	$1{\cdot}380 . 10^{-16}$	erg/°K
h	Planck's constant	$6{\cdot}625 . 10^{-34}$	J s	$6{\cdot}625 . 10^{-27}$	erg s
$R = kN$	Gas constant	$8{\cdot}314$	J/°K mol	$8{\cdot}314 . 10^{7}$	erg/°K mol
π		$3{\cdot}1416$	—	$3{\cdot}1416$	—
ϵ_0	Vacuum permittivity	$8{\cdot}827 . 10^{-12}$	farad/m	1	—
μ_0	Vacuum permeability	$1{\cdot}257 . 10^{-6}$	henry/m	1	—
hc/k		$1{\cdot}439 . 10^{-2}$	m °K	$1{\cdot}439$	cm °K
V_0	Normal volume of an ideal gas at 0°C and 1 atm. pressure	$2{\cdot}241 . 10^{-2}$	m^3/mol	$2{\cdot}241 . 10^{4}$	cm^3/mol

APPENDIX II

CONVERSION FACTORS

The equations in this book have all been written in SI units and, especially with electromagnetic expressions, correct results can only be obtained if all quantities are expressed in these units. Numerical values in the literature generally refer to Gaussian units and must be converted to rational units before the equations in this book can be applied. Moreover, spectrometric texts frequently use units of incorrect dimensions. For example, energy levels may be expressed in cm^{-1}, and magnetic field strengths may be given in gauss.

In the following table the = sign connects two units from different systems, but of similar dimensions. The $\simeq$ sign, however, must be interpreted as 'equivalent to'. Thus the expression x Å $\simeq$ 12·39/x keV means that radiation with a wavelength of x Ångström units has a photon energy of 12·39/x kilo-electronvolt.

$$
\begin{aligned}
1\ \text{cm} &= 10^{-2}\ \text{m}\\
1\ \text{g} &= 10^{-3}\ \text{kg}\\
1\ \text{dyne} &= 10^{-5}\ \text{newton (N)}\\
1\ \text{erg} &= 10^{-7}\ \text{joule (J)}\\
1\ \text{eV} &= 1{\cdot}602 . 10^{-19}\ \text{J}\\
1\ \text{eV} &\simeq 8068\ \text{cm}^{-1}\\
1\ \text{cm}^{-1} &\simeq 1{\cdot}239 . 10^{-4}\ \text{eV}\\
1\ \text{cm}^{-1} &\simeq 1{\cdot}985 . 10^{-23}\ \text{J}\\
x\ \text{Å} &\simeq 12{\cdot}39/x\ \text{keV}\\
x\ \text{nm} &\simeq 1239/x\ \text{eV}\\
x\ \mu\text{m} &\simeq 10^4/x\ \text{cm}^{-1}\\
x\ \text{cm}^{-1} &\simeq 10^4/x\ \mu\text{m}\\
1\ \text{e.s.u. of charge} &= 3{\cdot}336 . 10^{-10}\ \text{C}\\
1\ \text{gauss} &= 10^{-4}\ \text{V s/m}^2\ \text{(unit of magnetic induction)}\\
1\ \text{oersted} &= 79{\cdot}58\ \text{A/m (unit of magnetic field strength)}\\
1\ \text{gauss} &\simeq 1\ \text{oersted}\\
e^x &= 10^{0\cdot4343x}\\
10^x &= e^{2\cdot303x}
\end{aligned}
$$

ANSWERS

Chapter 1

1. Answer for KCl: 2·01 . 10^8 m/s; 395 nm; 28·3°
2. 1·54*A*
3. Answer for 20 μm: 1·5 . 10^{13} Hz; 500 cm^{-1}; 0·062 eV
4. $n_n/n_m = 5$
5. Sensitivity 136 V litre/mol; limit of detection 0·0066 millimol/litre
6. 0·0285 eV; 6·9 . 10^{12} Hz; 0·0435 mm; 750 and 250 particles

Chapter 2

2. 18 3. n = 3; l = 1, s = ½ 4. 17 cm^{-1}
5. 32700 cm^{-1} or 6·5 . 10^{-19} J; 1·13 . 10^{14} Hz; 785 N/m
6. 480 N/m; 2·2 cm^{-1}
7. (*a*) 0·0007 nm (*b*) 0·0043 nm (*c*) 0·0109 nm

Chapter 3

1. Answer for 2 nm/mm: 0·2 nm; 0·66 μm^{-1}
2. Answer for KCl: 0·0017 μm^{-1}; 0·0034 μm^{-1}; 0·06 μm
3. (*a*) 2 nm (*b*) 498 nm (*c*) 500 nm (*d*) linearly increasing from 498 to 500, linearly decreasing from 500 to 502
4. Answer for 500 nm: 37°; 18°
7. (*a*) B (*b*) B (*c*) A (*d*) B

Chapter 4

5. 0·635 mg; 6·25 per cent
6. *a* = 10600 l/mol cm; yes
9. 500 nm, red solution, green filter
   600 nm, blue solution, yellow filter
   700 nm, green solution, red filter
10. Curve (a), T_{min} = 37 per cent; curve (b), T_{min} = 13 per cent
11. The absorbtivities of B and C must be different

Chapter 5

3. 3000 cm^{-1}, 1700 cm^{-1}, 1400 cm^{-1}; no 4. No
5. 1700 cm^{-1} 9. See Chapter 8 11. Yes

Chapter 6

6. $I_1/I_2 = 5{\cdot}5$
8. (*a*) 378 and 535 nm

Chapter 7

1. 20 Ω **2.** $\gamma = 1{\cdot}45$ **4.** 0·030 per cent; 0·0025 per cent; no

Chapter 8

1. 36·8 μm **2.** 1·85 per cent **3.** $\theta = 11°$ (compare § 8.3)
4. 0·475 and 0·470 nm **6.** 11·4 per cent **7.** 17 μm; 53 per cent
8. 1·53 per cent **9.** About 3 minutes **10.** 77 counts; 0·025 per cent Al

Chapter 9

1. 8 and 6 **2.** 17; 0·185 per cent **3.** 0·78 MeV **7.** 0·75 per cent
8. 41 counts per second

Chapter 10

1. See Fig. 10.1 **2.** Na, 15·85; Cl, 5·87; V, 5·97; Cr, 0; In 13·12 MHz **5.** See Fig. 10.4c **9.** Propanol-2; 1 : 1 : 6

Chapter 11

1. 1880 **3.** 0·017 per cent **5.** C_2H_4 **7.** C_2H_5Cl **8.** 99·8 per cent ^{16}O, 0·2 per cent ^{18}O **9.** Benzene **10.** 7·7 **11.** Acetone

INDEX